Life and Lightning
The Good Things of Lightning

Life and Lightning
The Good Things of Lightning
1st Edition

Vernon L. Mangold, Ph.D.

Electromagnetic and Environmental Protection Consultants
Orlando, Florida

Universal Publishers/uPUBLISH.com
USA • 1999

Life and Lightning, 1st Edition.

Copyright © 1999 by Dr. Vernon L. Mangold

All rights reserved. Printed in the United States of America. No part of this book may be used or reproduced in any form or by any means, or stored in a database or retrieval system, without prior written permission of the publisher except in the case of brief quotations embodied in critical articles and reviews. Making copies of this book for any purpose other than you own personal use is a violation of United States copyright laws. For information, address EEP, 8121 Forest City Road, Orlando, FL 32810.

ISBN 1-5812-796-0

This book is sold as is, without warranty of any kind, either express or implied, respecting the contents of book, including but not limited to implied warranties for the book's quality, performance, merchantability, or fitness for any particular purpose. Neither EEP nor its dealers, distributors, or publishers shall be liable to the purchaser or any other person or entity with respect to liability, loss, or damage caused or alleged to be caused directly or indirectly as a result of this book or its contents.

Published by:
Universal Publishers/uPUBLISH.com
www.upublish.com/books/mangold.htm

Table of Contents

Forward 1

Introduction 2

Lightning – A Force For Life 5

Lightning Renews Life through Fire 9

Lightning Nutrients for a Living Earth 13

The Sound and Rhythm of Lightning 18

Man's Dependence in the Charged Atmosphere 22

    The Charging of the Atmosphere Begins With lightning 22

    The Creation and Maintenance of the Charged Atmosphere 26

    Lightning Induced Effects on Rain 31

    Harnessing the Electrical Energy in the Atmosphere. 33

    Atmospheric Electricity and its Effects on Pollution 33

    Lightning's Effect on Cleaning the Atmosphere of Sub-micron Particles. 35

Biological Affects Associated With Lightning 41

    Vegetation... 44

Reduction of Lightning Activity............................... 50

Conclusions... 55

Appendix A. Types of Lightning Discharges
And Flashes... 56

    Cloud-to-Ground Lightning............................ 60
    Intra-Cloud Lightning.................................... 60
    Cloud to Cloud Lightning.............................. 60
    Heat Lightning... 61
    Cloud-to-Air Lightning................................. 61
    Sheet Lightning... 61
    Ribbon Lightning... 62
    Bead Lightning.. 62
    Ball Lightning... 63
    St. Elmo's Fire.. 63

Appendix B. Lightning-like Discharges..................... 64

    Volcanic Induced Lightning........................... 64
    Nuclear Detonation Induced Lightning.............. 64
    Tornado/Waterspout Induced Lightning............. 65
    Dust/Sand Storm Induced Lightning................. 65

Appendix C. The Signature of Lightning..................... 70

Appendix D. Summary Guide for Personal Safety
During Thunderstorms. 73
.

Personal Safety During Thunderstorms 73

Be Watchful For Thunderstorm Activity 73

Cloud Formations Associated with
Thunderstorm Activity 74

Appendix E. 80

        Glossary... 80
        Units of Measure...................................... 86

Appendix F. Other Suggested Reading...................... 91

Foreword

This book was written to address those features of lightning that are beneficial to mankind. The technical level of the subject matter discussed in this book is designed to be readily understandable by the general public. The general emphasis is on the positive aspects of lightning. There are a number of books and thousands of published articles that stress the hazards of lightning to life and property. Lightning is an awesome force of nature and must be respected as such. I have included Appendix C; "A Summary Guide for Personnel Safety During Thunderstorms" which addresses simple steps one can take to help avoid this natural force. Other Appendices are included which discuss the types of Lightning discharges, Lightning-like Lightning, the Signature and color of lightning and other information to assist in the understanding of the phenomena of Lightning.

A large portion of the information used in this book is based on the results of other published work. Any attempt to write a comprehensive book on the subject of "Lightning" without due regard for the experiences and research of others in this technologic discipline would be impossible. The physics associated with Lightning, as we know it today, is based on the results of research and study conducted by many authors.

Lightning was part of the earth's environmental makeup before life appeared on earth. Lightning played an important role in creating life on earth. This same force has assisted in sustaining a livable earth and continues to balance those forces of nature to assure that life continues on this earth in the future. By studying such natural forces as lightning, we can better understand the workings of the past, present, and future of our natural planetary environment.

Introduction

A proposed theory, first reported in 1944, describes the creation of the universe as a result of electrical discharges (primordial-like Lightning) that gradually condensed the matter, consisting of gases and dust into the galaxies and finally into the stars. These lightning-like discharges ultimately condensed the stellar matter into planets and satellites. On our Earth, the electrical discharge phenomena (i.e. lightning), plays an active part in sustaining life and is one of those natural forces that has been around since the formation of the atmosphere.

Man can neither control nor regulate this force and resigns himself to the task of measurement and understanding of this phenomena. It is the purpose of this book to attempt to help one understand the importance of this great force as it was in the beginning of life, its role in nurturing life, and its ability to help sustain life on the earth.

As our earth was cooling down, our atmosphere was condensing from clouds that included small portions of heavy elements dispersed by the formation of the planet. With the oceans established, there was still a void in living matter - "LIFE." It is theorized and laboratory experiments suggest that lightning and ultraviolet radiation were responsible for the synthesis of the organic chemicals that eventually led to the development of life. Along with other natural processes, the primeval thunderstorm (lightning) helped produce molecular oxygen that would flourish in the oceans, on the land and in the atmosphere. Today, eons later, the thunderstorm with its associated lightning continues to nurture the life it helped create.

Lightning provided man with his first source of fire. This fire ultimately "blazed" the trails for the migration of man across the continents of the world. The lightning induced fires that cleared the forest and woodlands created the vast grasslands and thereby provided life and food for ancient man.

The brilliant lightning flashes generate tremendous energy and heat which unite the nitrogen and oxygen in the atmosphere to form nitrates and other compounds. These compounds are then carried to the earth by rain and replenish the supply of fertilizer that the soil needs to produce food.

The electromagnetic "sound" of lightning, which echoes back and forth between the earth's hemispheres, produces radio signals that, when amplified, sounds like the descending pitch of a whistle. By analyzing these signals, there exists a possible means to detect irregularities in the ionosphere and variations in the earth's magnetic field. This type of information could be used to study the upper atmospheric relationships with the sun.

A thunderstorm maintains the charged atmosphere which provides the electrical activity that helps keep the upper atmosphere free of sub-micron particles that could collect and create climatic changes by interfering with the solar energy.

Mankind lives in this electrically charged atmosphere and breathes in the charged air, which is absorbed in the respiratory system. The significance of this fact is that the current levels involved are too small to cause any noticeable effects. As humans, we have evolved and adapted to our natural environments; one of which is the charged atmosphere, which is sustained by thunderstorm activity around the earth. What about tapping into the electrical energy stored in the charged atmosphere for use in providing power for commercial and industrial uses? The electrical energy (lightning is the major source of this energy) stored in the atmosphere of the earth could supply millions of horsepower continuously if it could be harnessed.

The births of thunderstorms are a function of the surrounding atmosphere and the topography of the earth's surface. The global air circulation patterns control the planets weather as well as the development of the thunderstorm activity. The earth's air circulation patterns are a function of the motion of the earth and absorption of solar energy. The earth's atmosphere absorbs solar energy and the earth's motion helps establish the movement of the heated air creating a large atmospheric heat engine. The resulting energy builds and forms large cloud masses, high energy atmospheric winds and large atmospheric temperature gradients which pumps tremendous energy into the earth's surrounding atmosphere. One of the natural mechanisms for safely discharging this energy to the earth is through the development of the thunderstorm with its embedded lightning activity. Lightning is necessary to maintain the balance of energy within the earth's natural weather system. Lightning can be credited with performing a vital role in keeping the earth in electrical harmony with the upper atmosphere.

There are other attributes associated with this force of nature such as the production of ozone, which purifies the air we breathe after a thunderstorm. Lightning plays a role in accelerating rain and snow formation, the process by which energy is expended within weather related storm formations and finally the spectacular show provided to man representing the phenomena which helps nature balance the great forces which give life to our planet.

Lightning – A Force for Life

Creation of the earth began with a tremendous amount of energy being expended. This energy, among other things, created huge amount of heat (see *Figure 1*).

Figure 1. Primeval Earth Formation Showing Lightning Activity

As the earth cooled, its atmosphere could not hold all the water vapor that was in it. The vapor began to condense and great thunderstorms poured rain from the sky. These rains continued for years recreating the oceans and the first lightning.

Water, the universal solvent, formed into rain drops which washed carbon dioxide and other soluble compounds out of the air and carried them down to the earth and finally out to the sea. As the water passed over the earth it picked up other chemicals. Salts too were carried to the sea. These salts and other chemicals were mixed and reacted in the sea to form new chemical combinations which ultimately became the raw materials of life. Among the chemicals washed into the sea were carbon and nitrogen compounds that served as the raw materials from which living things developed. Some hydrocarbons and ammonia may have been left from the first atmosphere, volcanic eruptions were bringing molted rock and gases to the surface which reacted to form hydrocarbons and carbides. The carbides reacted with nitrogen to form cyanamides which reacted with water to produce ammonia.

Water, hydrocarbons, and ammonia are the raw materials out of which amino acids can be made. Two forces which existed at that time which could combine the raw materials into a form of living matter were ultraviolet light and lightning. Experiments in laboratories with electrical arcs which simulate lightning and ultraviolet energy, have proved that they are capable of altering the raw materials to form amino acids. Lightning can produce the synthesis for combining the raw materials to produce Amino-acids in two physical processes; heat from the lightning channel and the shock wave produced by the expanding channel which produces thunder. A large percentage of the electrical energy in a lightning discharge is dissipated in the lightning channel shockwave. The literature indicates that calculations show that ultraviolet light below ** 3000 Angstroms could be 1000 times more abundant than lightning on the primitive earth, but the shockwave from the lightning channel is a million times more efficient in producing Amino-acids - "The start of living matter."

The lightning as we know today was very surely one of the forces that helped create life as it is today. This great force that can destroy and start great forest fires was also one of the great forces that started life on our planet Earth.

** Wavelengths shorter than about 3000 Angstroms is where the organic photochemistry can be expected.

REFERENCES

FOX, S., "HOW DID LIFE BEGIN, "SCIENCETECHNOLOGY, FEB. 1968.

DAUVILLIER, A., "THE PHOTOCHEMICAL ORIGIN OF LIFE," ACADEMIC PRESS, NEW YORK, 1965.

OPARIN, A. I., "THE ORIGIN OF LIFE," MACMILLAN PUBLISHING, NEW YORK, 1938.

ADLER, IRVING, "HOW LIFE BEGAN," SIGNET SCIENCE LIBRARY BOOKS, NEW YORK, 1957.

"THE SEARCH FOR LIFE'S ORIGIN," NATIONAL ACADEMY PRESS, WASHINGTON, D.C., 1990.

MELVIN, CALVIN, "CHEMICAL EVOLUTION," OXFORD UNIVERSITY PRESS, NEW YORK, 1969.

BUVET, R. AND PONNAMPERUMA, C., "CHEMICAL EVOLUTION AND THE ORIGIN OF LIFE," AMERICAN ELSEVIER PUBLISHING COMPANY, INC., NEW YORK, 1971.

BRUCE, C.E.R., "AN ALL-ELECTRICAL UNIVERSE," ELECTRICAL REVIEW, LONDON, 167:1070-1075, 23 DECEMBER 1960.

Lightning Renews Life through Fire

Are lightning-produced fires mans great enemy or are they natures way of stabilizing life on earth? Did lightning "blaze" the trails for the migration of man across the continents of the world? It is postulated that the lightning induced fires were mans first source of fire which he used for heat, light, cooking, and protection.

The existence and finally the control of fire was a major step in man's evolution and mastery over his environment. As the thunderstorm built-up over the prehistoric forest, lightning ignited trees and grasses. When the fires burned themselves out; a small group of hunters moves into the smoldering areas. They find the carcass of a deer, and they taste it. They carry it home; food can be cooked! This miracle of fire was not limited to prehistoric man; an example situation was reported during the seventeenth century among the aborigines of Tasmania. Many of the tribes did not know how to create fire. They lived in an area where there were no active volcanoes. Their source of fire was obtained from forest and brush fires which were started by lightning. These people cherished and carefully nourished this most critical possession. On occasion, the fire was lost due to such occurrences as local flooding. When this happened, they would have to try to barrow fire from other tribes or they would have to go without until the next naturally produced lightning fire. These same lightning induced fires cleared the great forest and woodlands creating the grasslands which provided life and food for ancient man (*Figure 2*).

FIGURE 2. Fire was of critical importance to humanity; it provided warmth, protection and a means of cooking food.

In areas such as Central Arabia, the significance of lightning take on a different meaning: rain is eminent. Swift movement is essential to reach the water before it has evaporated or sunk into the ground. Lightning and its associated thunder herald the coming of rain showers.

The biologist Mr. Edwin V. Komarek, Sr. in his studies of lightning induced fire damage and the surviving ecology balance indicated that natures use of lightning fires for clearing dense wooded areas is indeed beneficial to the ecology. An example of an ecology out of balance was the great timber forest in Oregon's Tillamark. Prior to 1933 the serenely beautiful, hushed and somnolent forest was deserted by living creatures. A fire set by lightning wiped out 250,000 acres of virgin timber. Rather than destroying the ecology, the flames had in fact rejuvenated the environment by killing off the dry - land snail, a carrier of worms that infested the lungs and livers of the local mammalian population. The fire freed the forest of insects and disease. The bird and animal communities began to thrive in the clearings left by the fire. This is not a unique case in history but rather a typical situation that repeats itself in time, with lightning acting as the agent in plant and animal evolution. In contrast to the above, Nova Scotian Alder Thickets untouched by lightning induced fires, have biologically "died."

The effect on the evolution of plant life is reflected in past literature where it shows that new genes had appeared following a forest and grass fire, and that laboratory experiments with elevated temperatures doubled the number of chromosomes of some species. One set of fifteen buds exposed to high temperatures yielded twenty-one offspring, all with differing characteristics from the parent plant; seven had strange characteristics never seen before.

Lightning fires in Alaska and Canada regularly clear the nesting sites for the migratory birds, the development of grasslands due to lightning induced fires in forest areas encourages the reproduction of moose in Alaska, elk in the Rockies, deer in California and the Everglades, snowshoe rabbits in Wisconsin and Minnesota, prairie chickens in Texas and Louisiana, wild turkeys from the Carolinas through the Mississippi valley and ducks and geese along the southeast seaboard of the Atlantic.

REFERENCES

LEAR, JOHN, "LIGHTNING AS A SCULPTOR OF LIFE", SATURDAY REVIEW, 49:57-62, JUNE 4, 1966.

KESSLER, EDWIN, "THE THUNDERSTORM IN HUMAN AFFAIRS", UNIVERSITY OF OKLAHOMA PRESS, NORMAN, OK, 1983.

CLARK, GRAHAME AND PIGGOTT, STUART, "PREHISTORIC SOCIETIES", ALFRED A. KNOFF, INC., 1965.

WALLACE, RONALD L., "THOSE WHO HAVE VANISHED", THE DORSEY PRESS, CHICAGO, IL, 1983.

BLUMENSTOCK, DAVID I., "THE OCEAN OF AIR", RUTGERS UNIVERSITY PRESS, NEW BRUNSWICK, NJ, 1959.

Lightning Provides Nutrients for a Living Earth

Lightning helps man produce food and helps keep the plant life abundant by producing nitrogen in a form which can be used to fertilize the earth's soil.

Lightning is a natural short-lived high current electrical discharge which takes place in the atmosphere. This lightning and the associated corona (point discharges) are a source of the production of charged particles (i.e. ions) in the earth's atmosphere. Ions are electrical charged atoms and/or molecules that create an electrically conducting medium. Air ions are electrically charged air particles that have taken up an electrical polarity of either positive or negative.

As the lightning arc passes through the atmosphere it separates the nitrogen (N_2) into ionized nitrogen which unites with the oxygen to form nitrites (NO_2) and nitrates (NO_3). The source of these nitrites and nitrates are from the atmospheric gases. The composition of the gases that makes up our atmosphere is shown in *Table 1*.

The nitrogen molecules readily dissolve in water droplets and rainwater and are washed into the earth below. This natural cycle is shown in *Figure 3*. Each bolt of lightning produces a small amount of organic nitrogen. However, there are more than 8 million lightning flashes per day distributed across the earth which produces an sufficient amount of organic nitrogen. It is estimated that there is approximately 100 to 400 million tons of organic nitrogen produced by lightning storms per year. The world surface area is approximately 5×10^8 km^2. Therefore dividing this area into 100 to 400 million tons would yield nearly 2 to 8 pounds of organic nitrogen for every acre on the surface of the earth. While it is not assumed that all of the fixed atmospheric nitrogen is generated within the lightning flash, it is important to note that the thunderstorm systems, which includes lightning contributes significantly to the world's source of fixed nitrogen.

GAS	PERCENT OF ATMOSPHERE
Nitrogen	78%
Oxygen	21%
Argon	0.93%
Carbon Dioxide	0.04%
Helium, Neon, Hydrogen, Ozone	0.007%
Krypton, Xenon	0.02%

Table 1. Gaseous Composition of the Atmosphere

Every organism needs nitrogen to survive. It is found in all cell nuclei, constituting about 16 percent of the proteins, which are the building blocks of healthy plant and animal tissue. There are two forms of nitrogen; inorganic and organic. The inorganic nitrogen is an inert gas (N_2) which makes up three-fourths of the air we breath. Almost all the earth's nitrogen is in this form. Plants and animals can not assimilate inorganic nitrogen. The organic nitrogen is a rare form created by separating inorganic nitrogen (N_2) and then putting it back together with oxygen so that it can be assimilated by plants and animals. This process is called nitrogen fixation whereby nitrites (NO_3) are processed.

Organic nitrogen is vital for life and is in constant demand. The process of fixation requires a vast amount of energy. There are few organisms capable of meeting these high metabolic requirements. To further complicate the production of organic nitrogen is an entire group of de-nitrifying bacteria which try to undo what the nitrifying bacteria species have done.

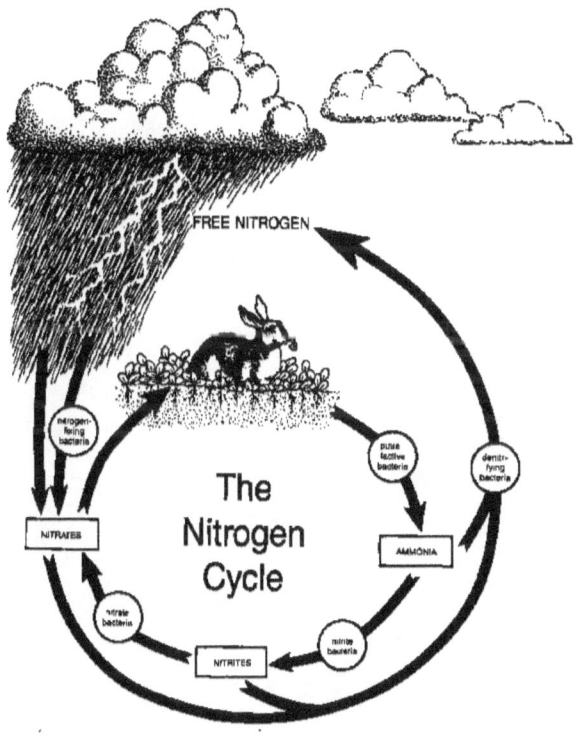

Figure 3. The Natural Nitrogen Cycle Produced By The Atmospheric Electricity.

This all means that either the forces of nature (thunderstorms) or artificial means must be used to supplement the nitrogen to the earth's soil. The artificial means is manmade nitrogen fertilizer, but this represents an energy loss, since plants grown from such fertilizers can never yield as much energy as was required to produce and transport the artificial fertilizers in the first place.

Fortunately, man has the thunderstorm with its embedded lightning working in his behalf as an important secondary source of organic nitrogen to help keep life as we know it sustained on our planet.

REFERENCES

LANGA, F.S., "FLASHES OF FERTILIZER IN THE SUMMER SKY," ORGANIC GARDING, 26:67-9, AUGUST 9, 1979.

STAFF,"LIGHTNING ACTS AS SOIL FERTILIZER," SCIENCE DIGEST, 11:47, APRIL, 1942.

STAFF,"LIGHTNING BENEFICIAL," SCIENTIFIC AMERICA, 172:52, JANUARY, 1945.

VIEMEISTER, P.E., "LIGHTNING AND THE ORIGIN OF NITRATES IN PRECIPITATION," JOURNAL OF METEOROLOGY, VOL. 17, NO. 6 DECEMBER 1960.

IRIBARNE,J.V.AND CHO, H.R., "ATMOSPHERIC PHYSICS," D. REIDEL PUBLISHING COMPANY, DORDRECHT, HOLLAND, 1980.

THE SOUND AND RHYTHM OF LIGHTNING

We are all aware of the tremendous sound (electromagnetic signal) associated with the electrical discharge of a cloud to ground lightning flash. A particular portion of that overall electromagnetic signal created by a lightning flash, echoes back and forth between the earth's hemispheres many times. These electromagnetic waves produce radio signals that when amplified, sound like the descending pitch of a whistle, whereby this phenomena obtained the name "Whistler(s)."

A Whistler can be described as an audio signal or band of noise lasting about one second in duration. The Whistling starts at a frequency of several kilocycles per second (kc/sec), then descends in pitch during the one second interval to a frequency of a few hundred cycles per second. The source of the audio signal is part of the very low frequency (VLF) band generated by the lightning return stroke. This sound of lightning is ducted through the ionosphere into the earth's magnetosphere along the lines of force of the earth's geomagnetic field and returned to the earth after a second penetration of the ionosphere. The auditory signal is altered "spread out" causing the signals of higher frequency to arrive first preceded by the lower-frequency signals resulting in a whistle with a distinct drop in pitch.

The spectral measurements of the whistlers can provide useful information associated with the outer ionosphere, or exosphere, of the earth. As an example, the path followed along the lines of the earth's magnetic field which leaves the northern hemisphere passes over the magnetic equator at a height of about two earth's radii, would make a round trip to the southern hemisphere and back. The total distance over which this round trip dispersion would take, would exceed 80,000 km. A conceptual drawing of the Whistler's path is shown in *Figure 4*. Concentrations and properties of the electrons and ions along the path of the electromagnetic signal affect both the modes of propagation and frequency of the auditory whistling signal.

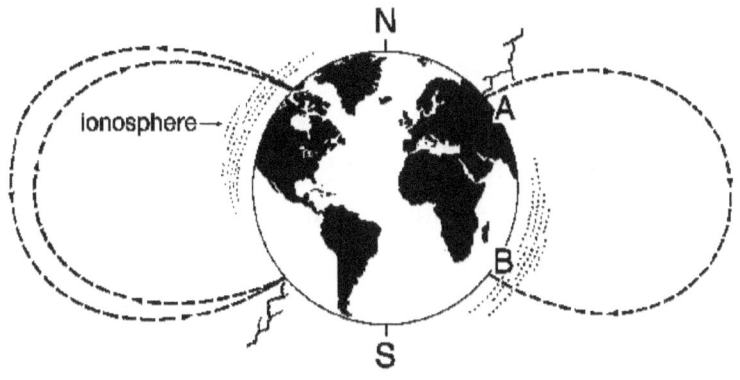

Figure 4. Whistler Path: A Lightning Signal At (A) Travels To Point (B) In The Opposite Hemisphere Undergoing A Dispersion On The Way, A Single-Hop Whistler Emerges At (B). A Reflected Signal May Be Returned To (A) To Be Heard As A Double-Hop Whistler

The sounds at the bottom of the radio spectrum are manifestations of planetary-scale processes driven by the lightning return flash and sustained by the geomagnetic field of the earth.

Other portions of the electromagnetic signal produced by lightning is trapped in a layer of earth's atmosphere called the troposphere in which weather systems are active. The conductive layer forms a spherical capacitor (described in more detail in Chapter VI) in which the low-frequency (7.9 Hz) electromagnetic waves are bounced around the globe. The fundamental mode of this resonance is equivalent to a standing wave which is equal to the circumference of the earth. This resonance field is called the Schumann resonance. The overall variation in the electromagnetic signals produced by the Schumann resonance can be measured at a single point on the earth's surface. These measurements could provide for real-time diagnostics of both the temperature and deep convection in the tropical atmosphere as measured from a single point on the earth's surface.

    The electromagnetic (sound) signal of the lightning flash could provide scientists around the world a means to study the ionosphere, magnetosphere, magnetic field effects, global temperature measurements and solar effects as associated with the earth and its atmosphere.

REFERENCES

WEISBURD, STEFI,"WHISTLING FOR LIGHTNING RHYTHM," SCIENCE NEWS, VOL. 131, Pages 372-3, JUNE 13, 1987.

MIDEKE, MIKE, "SPHERICS: A BEGINNER'S GUIDE TO WHISTLERS, TWIEKS AND OTHER RADIO SOUNDS AND HOW TO HEAR THEM," WHOLE EARTH REVIEW, Page 96, FALL 1990.

HUNTINGTON, CURTIS, "THE NATURE OF LIGHTNING DISCHARGES WHICH INITIATE WHISTLERS; RECENT ADVANCES IN ATMOSPHERIC ELECTRICITY," Pages 619-623, PERGAMON PRESS, NEW YORK, NEW YORK, 1977.

PIERCE, E.T., "ATMOSPHERICS AND RADIO NOISE," Pages 376-379, PHYSICS OF LIGHTNING, ACADEMIC PRESS, N.Y., N.Y. 1977.

VOLLAND, HANS," ATMOSPHERIC ELECTRO-NAMICS" SPRINGER-VERLAG,BERLIN, HEIDELBURG, NEW YORK, TOKYO, 1984.

MAN'S DEPENDENCE ON THE CHARGED ATMOSPHERE

To understand man's dependence on the charged atmosphere surrounding the earth, it is necessary to first describe the source of the energy that creates and maintains this charged atmosphere. It will be shown in this chapter that, the thunderstorm with its embedded lightning is that source of energy that sustains the charged atmosphere. With the charged atmosphere development and maintenance established, we are ready to examine the beneficial effects of the charged atmosphere, such as those effects on rain, atmospheric pollution and the cleansing of the atmosphere of sub-micron particles.

The Charging of the Atmosphere Begins With Lightning.

How does a thunderstorm develop and eventually produce lightning? The thunderstorm sequence begins with the rising of warm air, moist warm air near the earth's surface rises upwards and mixes with high-altitude cold air. The mixing causes the warm air to condense, forming clouds. As the process intensifies, the cloud formation grows and becomes very large. This gives birth to the giant Cumulonimbus cloud which spawns the thunderstorm. Within the cumulonimbus cloud formation is created violent updrafts, which when fully developed can produce winds exceeding 60 miles per hour. These updrafts within the cloud formation and the associated internal condensation produce and expand the clouds higher up into the sky. These cloud formulations can reach altitudes of 60,000 feet and higher where they meet up with the more stable Stratosphere. When the cloud mass meets the Stratosphere, the cloud top flattens and spreads out along the downwind course creating a giant anvil appearance which is characteristic of large and active thunderstorm cloud formulations.

These warm air updrafts are normally caused by three weather related conditions within the continental United States. The first being areas such as Florida and other Southeastern states that are near the sea, which produce drastic temperatures changes between the land and adjacent water. In the Midwest and east, these updrafts maybe caused by air that has been heated by radiation from the earth's surface and lastly; the updrafts that result from the movement of a cold front into a warm or hot environment. Of the three source producing the upward moving air, the sudden and rapid moving cold front results in the most violent and extensive development of the thunderstorm activity. The cold front produced storms are the most severe and are the longer lasting thunderstorms. The other produced thunderstorms normally are not as severe and last for just a few minutes up to about one (1) hour before they move out of the area.

The lightning action takes place in rapidly developing cloud mass. This cloud mass that spawns the thunderstorm is normally white from the mid portion-up and consists of a gray, to almost black at the cloud base. The cloud continues to grow in size as water vapor condenses and water drops coalesce. As the cloud builds upward into the colder air, the water droplets come into contact with ice crystals. As precipitation begins, the raindrops and ice crystals cool the warm air causing downdrafts alongside of the upward moving air. The colliding raindrops and ice crystals are continuously mixed together in the turbulent air. The friction of the continuing contacts create the static electricity build-up within the clouds, with a charge separation of positive and negative charged cloud portions. Normally, the lower portions of the cloud become negatively charged and the upward portions become positively charged. This charge steadily collects and produces a very high intensity charge separation within the cloud. When the voltage level exceeds the breakdown resistance of the surrounding air (i.e., electrical flashover); the process for the development of lightning begins. The subsequent lightning flashes can take a course of cloud to cloud, within the cloud or between the earth and sky. Normally 65% of all the lightning flashes are intra-cloud or within the cloud mass itself. No matter where the lightning exists, it always tends to travel towards the earth due to the electrical attraction of the earth. As a result of this electrical attraction of the earth, the developed charge centers within the clouds initiate and sustain a ionized path along which electricity can be conducted between the earth and the cloud mass. The lightning flash is composed of a number of lightning strokes that repeat the lightning event normally using the same initial lightning stroke path.

There is first, a downward leader that begins its travel to the earth from the base of the cloud producing a narrow ionized air channel behind it. As this leader nears the earth, a stream of ions from the nearest earthbound object/feature, leaves the earth and joins the downward ionized leader forming the lightning return stroke. It is the return stroke that we see as lightning. The extent of the energy developed in the lightning stroke is summarized in *Table 2*.

Thunder is the secondary phenomena of the lightning event. It is born the instant the return stroke is created. The sound is produced by an explosion that occurs along the entire length of the lightning channel. The lightning return stroke develops the greatest energy during the lightning flash and produces the loudest "sonic boom." This return stroke raises the temperature of the air within the lightning channel to temperatures in excess of 30,000 degrees centigrade in such a short time that the air within the channel increases almost instantaneously to 10 to 100 atmospheres. The high pressure channel expands rapidly and radially out into the surrounding air. This produces a very powerful shock wave (which travels faster than the speed of sound) and farther out from the channel, the shock wave is heard as thunder. Thunder is also produced by stepped and dart leaders, but is much weaker than that produced by the return stroke.

ELECTRICAL

TYPES: - Intra/inter - cloud, cloud to ground, positive, negative.
POTENTIAL: - 30 to 100 million volts
CURRENT: - 20 to 200 thousand amperes
POWER: - 100 trillion watts
ENERGY: - 0.5 billion Joules (normal)
            (200 lbs. of TNT equivalent)
CHANNEL TEMPERATURE: - 54,000 degrees F.
EXTENT: - 3 to 30 Km per stroke
SPECTRUM: - peak energy near 10 kHz, some above 10 MHz.
DURATION: - Stroke - 100 microseconds
            Flash - 0.2 Seconds (1 - 20 strokes)
OCCURRENCE / EFFORTS:
 	
Worldwide phenomena; 100 flashes per second average; activity varies with climate, season, hour, location, altitude, turbulence correlated with lightning activity.

Table 2. Lightning Characteristics

The sound originating from the lower portion of the channel is heard by observers relatively close to the lightning flash, while the sound of the upper portion of the channel is lost to the observers who are further than 10 to 15 miles away. At these distances, the thunder passes over the observers head because the sound is generally refracted (bent) upwards up temperature gradients which exist during the thunderstorm.

Generally, the duration of the thunder is determined by the time required for sound originating in the upper portion of the lightning stroke to reach the listener. Sound can be reflected from cloud surfaces near the top of the lightning channel which can travel parallel to the surface of the earth creating long duration thunder reports. Where these sound waves are reflected and bent in different ways due to the nature of the thunderstorm conditions; the resulting characteristics is that the thunder will be heard as a reverberating sound that at times echoes and re-echoes across the sky.

As awesome and frightening as lightning and the associated thunder can be at times, this is one of most commonplace weather phenomena that mankind can experience, the same phenomena which existed at the beginning of life is repeated continuously around the earth.

The Creation And Maintenance Of The Charged Atmosphere

The atmosphere is a deep ocean of gases, suspended solids and liquids that entirely surround the earth. The vertical extent of this atmosphere is difficult to determine as there is no sharp division between the air and the extraterrestrial space. For our discussion, we have divided the atmosphere into three major layers; the Troposphere, Stratosphere and the Ionosphere as depicted in *Figure 4*. The Troposphere contains three fourths of the atmospheric mass. This layer is associated with our weather (clouds, storms and convective motions). The next layer is the Stratosphere which has the greatest concentration of ozone and the occasional appearance of Nacreous (Mother-of-Pearl) clouds. The Ionosphere is a region of predominately charged particles (i.e. ions). There are so many ions and free electrons in this layer that this region is an excellent conductor of electricity. The ionosphere reflects downward the lightning produced radio waves.

The creation and maintenance of the charged atmosphere surrounding the earth is a direct result of thunderstorm activity. The electrical atmosphere can be visualized as shown in *Figure 5*, as a huge concentric - spherical capacitor.

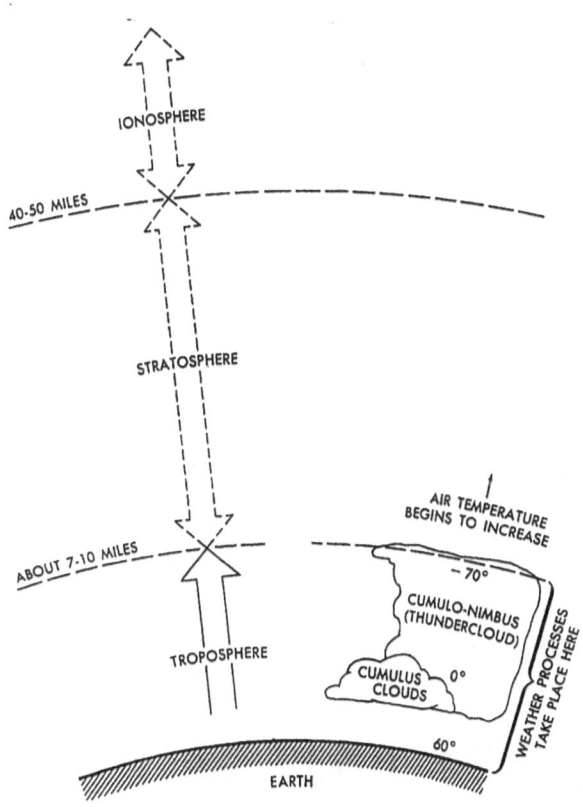

Figure 5. The Atmospheric Layers Above the Earth Showing the Height of the Ionosphere.

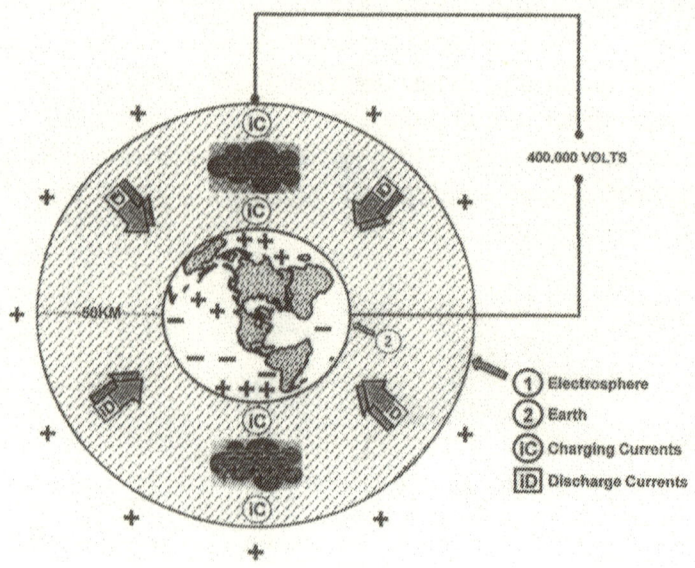

Figure 6. Spherical Capacitor Representation of the Atmospheric Electrical Environment.

Within the ionosphere, there is a imaginary region called the electrosphere which represents a good electrical conducting region "surface" which serves as the positive plate of the spherical the capacitor. One plate of this capacitor would be the earth's surface, the other plate of this capacitor is the electrosphere. The ionosphere is a better electrical conductor than the bottom of the electrosphere region. The electrosphere, located some 50km above the earth, is a region where ionization caused primarily by cosmic rays has created sufficient ions and free electrons to make the surface (area) a good conductor. The electrosphere is normally considered the height of the charged atmosphere as represented in *Figure 6*. The fair-weather potential (voltage) extends up to an altitude of approximately 40 to 50 miles. The dielectric in this large capacitor is the atmosphere which is composed of air, water vapor, dust and other gaseous and solid matter that nature and/or man has deposited in the atmosphere. This capacitor is not a very good one because the atmosphere is a poor dielectric and does conduct electricity. Charges move from one plate to another in response to fields created by potential differences between plates. The rate of flow, called the air-earth or ionic current, has been calculated to be 1400 to 1800 amperes. The ionic current flows upwards in fair weather from the surface of the earth. This current is too small to be felt, amounting to only approximately .000009 amperes for every square mile. This difference in potential between the earth and the ionosphere ranges between 300,000 to 400,000 volts. The "fair-weather" electrical field intensity that is created, measured near the ground is approximately 100 to 300 volts per meter and negative; i.e. the earth is negatively charged and the atmosphere above the earth is positively charged. The opposite is true during cloud cover as shown in *Figure 7*. The term fair-weather is used to mean without clouds, the term fine weather is sometimes used interchangeably. The value of the fair-weather charge varies with time and location. These variations are caused by the amounts of particulate matter in the atmosphere (i.e. dust, aerosols, salt particles, etc.), atmospheric humidity and the surrounding thunderstorm conditions. It is interesting to note that for a six foot tall man; his head is at potential of about 300 volts as compared to the bottom of his feet (see *Figure 8*). The fair-weather potential gradient varies from month to month reaching a maximum of 20 percent above normal in January when the earth is closest to sun and 20 percent below normal by July when the earth is farthest from the sun.

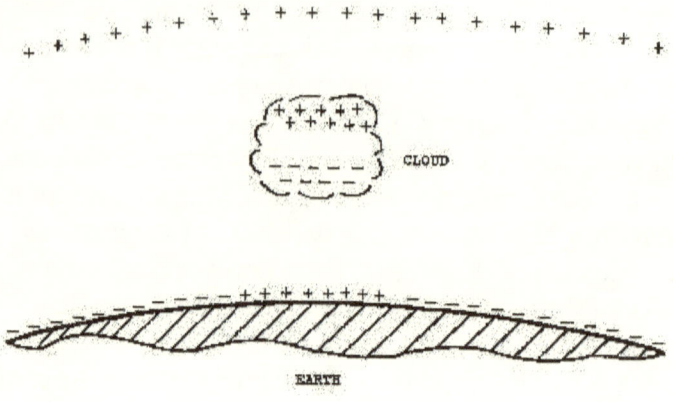

Figure 7. Atmospheric Electricity Charge Orientation.

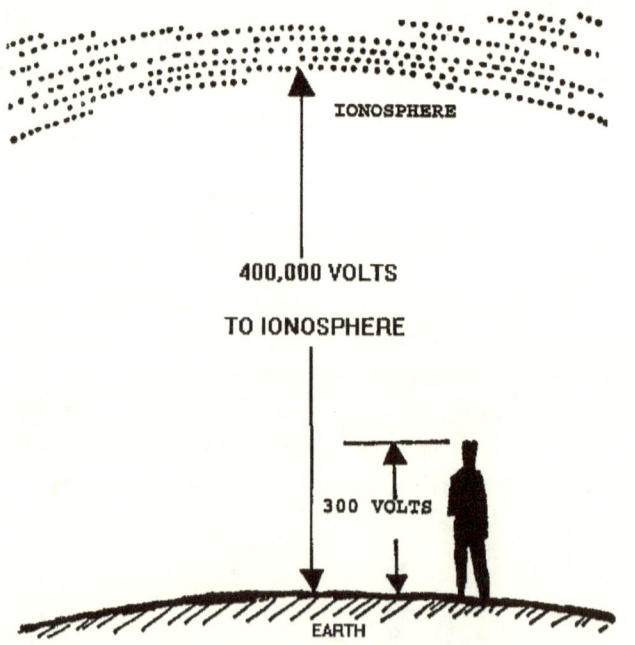

Figure 8. The Potential Gradient that Exists due to the Voltage difference between the Earth and the Ionosphere

Lightning Induced Effects On Rain.

The Earth's fair weather electric field which is maintained by lightning activity around the world initiates and assists in sustaining the polarization of rain drops within a cloud mass. The interaction of particles, particularly collisions and separations, in the existing fair weather electric field with potentials as small as one volt per centimeter must be taken into account when computing the growth of the electric field within clouds. The rain drop polarization further promotes the separation of charge within a cloud mass which in-turn increases the electric field within the cloud mass. The electric field is further increased and maintained by the electrical charge transfer between water drops and or ice crystal as they come into contact with each other. The smaller moisture droplets, through collisions transfer, coalesce those drops to form larger water droplets, leaving the bigger drops increasingly negative as they fall and the smaller droplets increasingly positive as they move upwards. The cloud mass continues to charge by the collisions of water droplets and ice crystals which are kept in motion by the upper atmospheric winds, thermal updrafts, natural diffusion and gravity. This continuing charging of the cloud mass promotes larger magnitude electric fields. When the charge within the cloud mass exceed the dielectric strength of the surrounding air; the phenomena of lightning is created.

The earth's fair weather electric field acts as the catalysis in promoting charge separation and electrical field build-up in a cloud mass. Without this fair weather electrical field, it is theorized that the earth would experience increased cloud cover, longer durations of that cloud cover and more sever thunderstorms. Rainfall patterns could change in both terrestrial location and rainfall amounts. This fair weather charging of the atmosphere is continuously being replenished by the thunderstorm activity around the world. The number of thunderstorms in progress at any one time around the earth has been variously estimated as between 1400 to 3000.

It has been observed that lightning and the subsequent acoustics associated with the resulting thunder coalesce the smaller water droplets into larger droplets and is followed by intense (gushes) rainfall. Not all rainfall is associated with lightning, but a larger percentage of rain storms have lightning associated with them. The buildup and establishment of rain producing cloud formations with the subsequent rainfall back to earth is an awesome natural atmospheric phenomena. To make one inch of rain over the state of Florida would take nearly four billion tons of water.

Florida has an average rainfall of more than 50 inches per year. With that figure in mind, imagine how much water is sucked up from the oceans, seas, lakes and rivers and dropped back again as rain or snow on the entire earth each year. It is estimated that rain or snow is always falling on the earth at a rate of 16 million tons per second. The estimated rainfall average for the earth is 40 inches per year. Nearly all of this rainfall is associated with rain storms with imbedded lightning activity (i.e., Thunderstorms). Using the value of 40 inches of rainfall per year would yield a total of 8×10^{11} tons of water falling on the earth per day. Since the land mass comprises approximately one-quarter of the earth's surface; the amount of water that can be used for human consumption would be 2×10^{11} tons per day. If this amount is divided by the earth's population, it would amount to 8 tons or more than 20,000 gallons per day for every human being.

Harnessing The Electrical Energy In The Atmosphere

The design and utilization of electrostatic devices powered by the natural occurring fair-weather environment offer unique challenges in various scientific and technologic areas. The electrical energy stored in the atmosphere surrounding the earth could supply millions of horsepower continuously if it could be harnessed. The obvious problem is in the collecting of the electrical energy which is spread over a very large area with the resulting consequence of this being that the energy density is very low. To collect the atmospheric electricity requires very large area-wise antenna collection system.

Some years ago in Switzerland, a wire screen several hundred square yards in area was suspended from insulated cables which was used to collect atmospheric electricity. The result was said to be an almost steady discharge of sparks 15 feet long from its terminals. It was maintained that with modifications, the apparatus could supply 30 million volts. This high voltage would have to be reduced and the current levels increased to be consistent with normal electrical power use.

Atmospheric Electricity and Its Effect on Pollution.

The sun and stars emit a full spectrum of radiation. A portion of that spectrum is high-energy cosmic rays. These cosmic rays continuously bombard the earth's atmosphere producing charged particles (i.e. ions) within the atmosphere. The cosmic rays dissipate the majority of their energy in the upper atmosphere, (the ionosphere) by charge transfer to the air molecules. The cosmic rays strike the air molecules stripping off electrons creating the ions. Ions are atoms or molecules that are not in electrical balance. When an electron is stripped off; a positive ion is produced. When an extra electron is picked up by a atom or molecule; a negative ion is produced. Ionized molecules can cluster together forming "small" ions particles/aerosols (i.e. sub-micron to 1 micron in diameter, see *Figure 9*). When these small ion particles/aerosols adhere to particles of matter such as sea spray, automobile exhaust particles, smoke particles from power plants; we have "large" ions or aerosol particles which

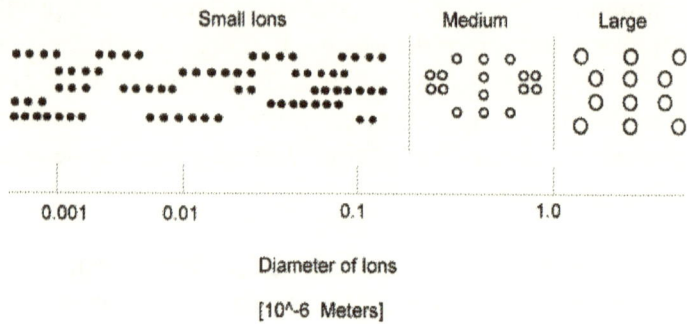

Figure 9. Ions Can Be Conveniently Classified In Three Size Groups

further combine together to produce atmospheric smog. Cosmic rays are only one source in producing ions in the earth's atmosphere. Nuclear radiation from the earth, lightning, and corona are other sources of additional ionization in the lower atmosphere, the troposphere.

Cloud droplets and ice crystals are formed around aerosol particles which act as condensation or freezing nuclei. When these droplets and ice crystals evolve into raindrops and snow flakes which fall to the earth, they "clean" the atmosphere of suspended nuclei or aerosol pollution. A major force which assist in uniting these aerosol pollution particles with the cloud and precipitation particles is the electrical environment present in the troposphere. This electrical environment (field) is associated with the fair weather field which is much larger in thunderstorm situations. When thunderstorm conditions prevail, the effects of removal of pollution from the atmosphere are greatly enhanced. Thunderstorms and its associated lightning activity acts to clean the atmosphere of man-made and natural pollution.

Lightning's Effect on Cleaning the Atmosphere of Sub-Micron Particles

As discussed above the fair-weather charged atmosphere which continuously transfers charge between the earth and the ionosphere. If the thunderstorms were not available to continue this charging sequence, the atmosphere would discharge itself to the earth within approximately one hour. This charged atmosphere is responsible for controlling and cleaning the upper atmosphere of sub-micron particles. The atmospheric particle/ion sizes where discussed in the previous section (see *Figure 8*). Particles in the micron to 10's of micron size decrease rather rapidly with increasing altitude and this is generally attributed to gravitational fallout. These large nuclei particles are also windblown and rapidly removed by skidding across aerodynamic streamlines and striking surfaces and other particles which remove energy and helps gravity pull them to the earth. Extremely fine particles (1 micron and larger) are quickly removed by diffusing to and attacking an obstacle under the driving action of Brownian molecular bombardment. This leaves the particles in the sub-micron size free to circulate and build-up to high levels in the upper atmosphere. These particles are relatively immune to removal and remain for long periods of time. The result of this build-up can produce what is termed the greenhouse effect. The sub-micron particles strongly interact with sunlight. The collective build-up of sub-micron particles/aerosols therefore constitute an enormously sensitive climate regulating machine.

In man's quest for food, goods, heat for homes and energy for our industrial society, we have increased the concentration of many trace gases and other pollution in the atmosphere. Couple this with the naturally produced sub-micron particles biologically produced sulfate aerosols, CO_2 and Nitrogen gases and other aerosols that have existed even before man, the upper atmosphere is becoming crowded. Some of these particles absorb solar energy and some have the effect to block solar radiation due to reflection. The sub-micron particles resonate on the same order as the wavelength of light to scatter, absorb, and reflect the solar radiation. The existence of these sub-micron sized particles create a potential for causing the earth's climate to change.

It was not until 1982 that the hypothesis of global cooling from soot, generated by nuclear war, could effect or change the climate, was taken very seriously. The concept since has been termed "Nuclear Winter". Even with the concept of limited nuclear war, it has been predicted that this would cause the generation and injection into the atmosphere of so much particulate matter that crops would freeze in the summertime.

Even without this man-made threat, there is the threat of nature which supplies to the atmosphere those sub-micron particles that, in a collective form, could produce changes in the earth's climate. The charged atmosphere has prevented the collection of sub-micron particles in the upper atmosphere that would have the effect of altering the earth's climate.

It can be shown that sub-micron particles (particles radius of 10^{-6} cm) carrying one elementary charge will experience a force, produced by the fair-weather electric field in the stratosphere, that is roughly 10 times as large as the gravitational force.

During thunderstorm activity, the relative electrical force will be proportionally much greater. This indicates that thunderstorms with associated lightning produces and sustains the charged atmosphere and imparts to the sub-micron particles, the electrical motion which acts to precipitate out these fine particles onto cloud droplets which subsequently fall back to the earth. If it were not for the charged atmosphere, these sub-micron particles would collect and stay suspended in our upper atmosphere, producing climate changes which would eventually spell the end of life on earth.

REFERENCES

FRISKEN, WILLIAM R., "THE ATMOSPHERIC ENVIRONMENT", THE JOHNS HOPKINS UNIVERSITY PRESS, BALTIMORE, MD, 1973.

WANG, P.K., "THE INFLUENCE OF ATMOSPHERIC ELECTRICITY ON THE PRECIPITATION SCAVENGING OF AEROSOL PARTICLES", 1986

PHILLIPS B.B., etal, "AN EXPERIMENTAL ANALYSIS OF THE EFFECTS OF AIR POLLUTION ON THE CONDUCTIVITY AND ION BALANCE OF THE ATMOSPHERE", Pages 289-296, 1955

GRIFFING, GEORGE W. "OZONE AND OXIDES OF NITROGEN PRODUCTION DURING THUNDERSTORMS", JOURNAL OF GEOPHYSICAL RESEARCH, VOL. 82, NO. 6, Pages 943-950, FEBRUARY 20, 1977.

TALUKDAR, R., "ATMOSPHERIC LIFETIME OF CHF2Br, A PROPOSED SUBSTITUTE FOR HALON" SCIENCE, VOL. 252, Page 693 - 694, 3 MAY 1991.

LEVINE, JOEL S., etal, "TROPOSPHERIC SOURCES OF NOx: LIGHTNING AND BIOLOGY", ATMOS-PHERIC ENVIRONMENT, VOL. 19, NO. 9, Pages 1797 - 1804, 1984.

WENT, FRITS W., "AIR POLLUTION", SCIENTIFIC AMERICA, VOL. 192:62+, MAY 1955.

FISH, BIRNEY R.,"ELECTRICAL GENERATION OF NATURAL AEROSOLS FROM VEGETATION", SCIENCE VOL.175, Pages 1239-1240, 17 MARCH 1972.

DICKERSON, R.R., "THUNDERSTORMS: AN IMPORTANT MECHANISM IN THE TRANSPORT OF AIR POLLUTANTS", SCIENCE, VOL. 235, Pages 460-464, 23 JANUARY, 1987.

SARTOR, J.D., "THE ROLE OF PARTICLE INTERACTIONS IN THE DISTRIBUTION OF ELECTRICITY IN THUNDERSTORMS", JOURNAL OF THE ATMOSPHERIC SCIENCES, VOL. 24, NO. 6, Pages 601-615, NOVEMBER 1967.

KUETTNER, JOACHIM P., "THUNDERSTORM ELECTRIFICATION-INDUCTIVE OR NON-INDUCTIVE?", JOURNAL OF THE ATMOSPHERIC SCIENCES, VOL. 38. Pages 2470-2484, NOVEMBER, 1981.

BOHREN, CRAIG, "THE GREENHOUSE EFFECT: PART III", WEATHERWISE, Pages 106-109, APRIL 1985.

GUNN, ROSS, "THE ELECTRIFICATION OF PRECIPITATION AND THUNDERSTORMS", PROCEEDINGS OF THE IRE, Pages 1331-1358, OCTOBER 1957.

DOLAN, EDWARD F., "BOOK OF WEATHER LORE", YANKEE PUBLISHING INCORPORATION, DUBLIN, NEW HAMPSHIRE, 1988.

CORN, PHILLIP B., "AN OVERVIEW OF LIGHTNING HAZARDS TO AIRCRAFT", PRIVATE COMMUNICATION, 1977.

ORVILLE, RICHARD E., "THE LIGHTNING DISCHARGE", THE PHYSICS TEACHER, Pages 7-13, JANUARY, 1976.

DAY, JOHN A., "THE SCIENCE OF WEATHER", ADDISON-WESLEY PUBLISHING CO., 1966.

BLUMENSTOCK, DAVID I., "THE OCEAN OF AIR", RUTGERS UNIVERSITY PRESS, NEW BRUNSWICK, N.J., 1959.

JEFIMENKO, OLEG, "OPERATION OF ELECTRIC MOTORS FROM ATMOSPHERIC ELECTRIC FIELD", AMERICAN JOURNAL OF PHYSICS, Vol.39, Page 776, JULY 1971.

JEFIMENKO, OLEG, "ELECTROSTATIC MOTORS", PHYSICS TEACHER, Vol 9, Page 121, 1971.

CHALMERS, J. ALAN, "ATMOSPHERIC ELECTRICITY", PERGAMON PRESS, LONDON, 1967.

KRAAKEVIK, J.H., "ELECTRICAL CONDUCTION AND CONVECTION CURRENTS IN THE TROPOSPHERE", RECENT ADVANCES IN ATMOSPHERIC ELECTRICITY, PERGAMON PRESS INC., NEW YORK, 1958.

Biological Effects Associated With Lightning

Humans are dependent on the electrical voltages and currents that are generated by the myriads of cells within his body. This effect can be termed bioelectricity. Three of the most common bioelectricity techniques used to diagnose humans are: electrical signals from the brain, which are measured by the Electroencephalograph (EEG); electrical signals from the muscles, which are measured by the Electromyography (EMG); and the electrical signals from the heart, which are measured by the electrocardiograph (ECG). Mankind is a creature of their environment being surrounded (see *Figure 10*) by a charged atmosphere since life began. This same charged atmosphere that was in part created by and now maintained by the phenomena of lightning.

There is ongoing interest in electric phenomena of the atmosphere and its influences on humans and their natural environment is noted. Many studies concerned with air ionization and its effects on the human body have been conducted. Air ions are electrical charged air particles. They can be oxygen atoms or water molecules that have taken up an electrical charge, either positive or negative. Studies have indicated that the positive ions may elicit unpleasant side effects within us, whereas negative ions counteract and abolish these unpleasant reactions. Studies show that negative ions promote the germination and subsequent shoot growth in vegetation. Ions are greatly increased in numbers during thunderstorm activity primarily through lightning discharges and ionization from strong fields within the thunderclouds. Most lightning discharges bring down negative charge to the earth from the clouds above.

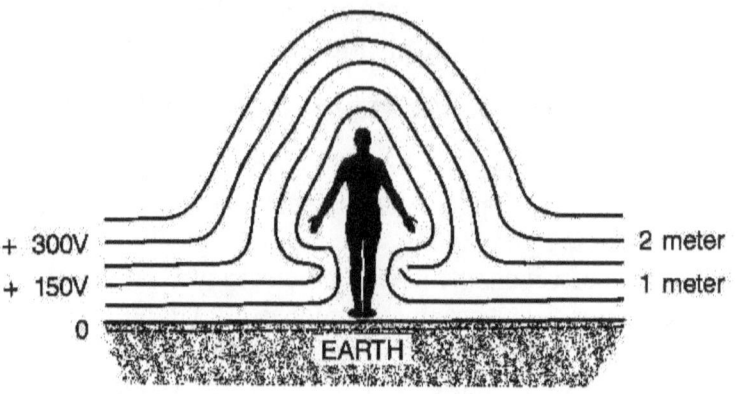

Figure 10. A Person Exposed To The Fairweather electrical field.

Hours to days before the arrival of an electrical thunderstorm, the air is overloaded with positive ions. With the passage of the electrical thunderstorm there is an overdose of negative ions which when inhaled and trapped in the respiratory system evoke a wide range of beneficial physiological and biochemical effects. All of the investigators associated with this area of research agree that negative ions and the atmospheric electric field in general are beneficial to humans and animals. A fluctuating electric field strength which is typical for thunderstorm activity promotes a tranquilizing, "feeling-well" sense in humans.

Electrical field magnitudes remain constant in time during fair weather conditions. The earth's fair-weather natural electrical field produces a negative charge at the surface of the earth and a positive charged upper atmosphere (see *Figure 7*). With the approach of clouds in the upper atmosphere, the polarity changes. The atmosphere adjacent to the earth is now positively charged and the upper atmosphere is negatively charged. The charge values near the earth during fair-weather has an average value of -3×10^{-12}

amperes/meter^2 while during cloudy conditions the current value near the earth's surface averages about $+4 \times 10^{-12}$ amperes/meter^2. These are very small current levels, these fair-weather electrical fields are too small to directly cause any detrimental effects to the human body. The possibility of interaction of living organisms with the natural atmospheric electric field(s) is represented only by indirect evidence which consists in-part by the fact that living organisms create internal weak electrical fields of various frequencies and are highly sensitive to those fields. Research studies have found evidence that electric potentials are associated with all living things. Small direct currents have been measured on plants, trees, insects and mammals. When addressing the effects on biological systems of the natural electrical and magnetic fields, certain biological mechanisms tend to respond to these weak natural occurring environments. The biological clock and biological compass of living organisms tend to respond and align themselves relative to the earth's electric and magnetic fields. Organisms possess a biological clock which coordinates the rhythm of biological processes with periodic changes in the environment. This response to the weak natural fields follow very closely and react with other environmental and geophysical inputs. Life has evolved over time in this charged atmosphere and appears to be very sensitive to changes in either the concentration or charge polarity.

A. Vegetation

There is a continuous flow of charge (electrons) between the earth and the atmosphere, both during fair weather and thunderstorm activity. Electrons flow from the earth to the atmosphere through sharp points. This includes electron flow from the tops of buildings, trees, blades of grass and even from living beings during fair weather. This flow of electrons greatly increase but reverse the flow from the atmosphere to the earth when thunderclouds are overhead. These "point-discharge's" are generally silent and invisible. The build-up of the electrical field near a point causes a spontaneous emittance of ions which, further ionize the air in the immediate vicinity by collision with gaseous molecules.

Figure 11 shows Dr. B. F.J. Schonland's point-discharge tree experiment in which the flow of electrons from the tree to the atmosphere was measured. This natural charge exchange has taken place since the development of the charged atmosphere. Mankind, as well as the earth's vegetation has never experienced life on earth without this flow of electrical charge.

An interesting part of lightning/vegetation interaction is that (has been suggested in the open literature), the organic vapors (haze) arising from vegetation due to the passing overhead of electrified clouds (i.e., thunderstorm activity) can establish a natural process for producing petroleum. Most of the aromatic substances emitted by flowers due to thunderstorm activity are hydrocarbons (i.e., Terpenes) or slightly oxidized hydrocarbons belonging to the general group of essential oils. There is hardly no type of vegetation which does not emit volatile organic compounds due to electrical atomization (i.e. corona) from overhead electrical storm gradients. The organic vapors coalesce and grow in size and ultimately become heavy enough to slowly settle

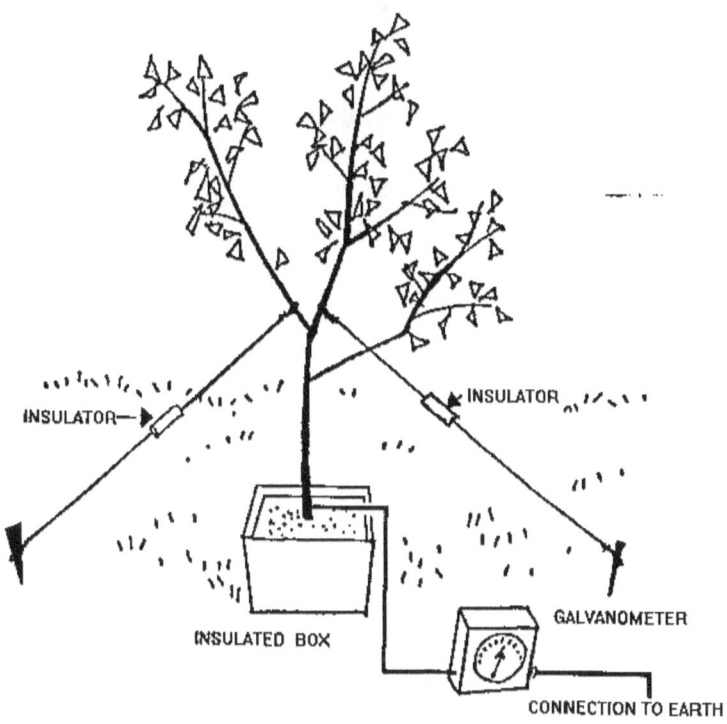

Figure 11. B.F.J. Schonland's tree experiment showed electrons leave the air, enter points on trees, and flow to earth when thunderclouds are overhead. Dr. also found that electrons pass from the earth, through the tree, and out into the air during fair-weather, but with less magnitude

back down to the earth. Chemical analysis of some rain and snow residue reveals the presents of organic material of black tarlike consistency. This haze (organic vapors) is normally bluish in color which is characterized as summer heat haze. This summer heat haze is a naturally produced smog. This smog (i.e., aerosols) is derived from vegetation due to the potential gradient produced by the charged atmosphere associated with thunderstorm type activity. The aerosols that make up the blue haze return to the ground by various natural processes. This resulting deposition could be the source of petroleum found in the early geological periods. The amount of organic material given off by the plant life would be sufficient to account for all the petroleum formed in previous geological periods.

REFERENCES

BATTOCLETTI, JOSEPH H., "ELECTROMAGNETISM, MAN AND THE ENVIRONMENT," WESTVIEW PRESS, BOULDER, COLORADO, 1976.

BECKER, ROBERT O.,SELDEN, GARY, "THE BODY ELECTRIC," WILLIAM MORROW AND COMPANY, INC., NEW YORK, 1985.

WAHLIN, LARS, "ATMOSPHERIC ELECTROSTATICS," RESEARCH STUDIES PRESS LTD, 1986.

THOMSON, K. S., "THE SENSE OF DISCOVERY AND VICE-VERSA," AMERICAN SCIENTIFIC, VOL. 71, Pages 522-524, 1983.

LANZEROTTI, LOUIS J. & GREGORI, GIOVANNI, GREGORI, P., "TELLURIC CURRENTS: THE NATURAL ENVIRONMENT AND INTERACTIONS WITH MAN-MADE SYSTEMS," NATIONAL ACADEMY PRESS, WASHINGTON, D.C., 1986.

McCURK, F. C. J., "PSYCHOLOGICAL EFFECTS OF ARTIFICIALLY PRODUCED AIR IONS," AMERICAN JOURNAL PHYS. MED., 38, 4, Pages 136-7, 1959.

GOLDE, R. H, "LIGHTNING," ACADEMIC PRESS, INC., NEW YORK, NEW YORK, 1977.

TROMP, S. W., "BIOMETEOROLOGY," THE MACMILLAN COMPANY, NEW YORK, 1962

KESSLER,E.&WHITE,G.F.,"THUNDERSTORMS IN A SOCIAL CONTEXT," NOAA-82032206, NOAA, ENVIRONMENTAL RESEARCH LABORATORIES, 1981.

OTT, JOHN N., "LIGHT, RADIATION, AND YOU," DEVIN-ADAIR, PUBLISHERS, GREENWICH, CONNECTICUT.

FOX, S., "HOW DID LIFE BEGIN!" SCIENCE & TECHNOLOGY, Feb., 1968.

CARSTENSEN, EDWIN L., "BIOLOGICAL EFFECTS OF TRANSMISSION LINE FIELDS," ELSEVIER, NEW YORK, 1987.

SOYKA, FRED, "THE ION EFFECT," E.P. DUTTON & CO., INC., NEW YORK, 1977.

KESSLER, EDWIN, "THE THUNDERSTORM IN HUMAN AFFAIRS," UNIVERSITY OF OKLAHOMA PRESS, 1983.

KONOW, R.V., "THE THUNDERSTORM AS A CHEMICAL PHENOMENON," JOURNAL OF THE FRANKLIN INSTITUTE, VOL. 269, No. 6, Pages 439-444, June 1960.

LATHAM, J., "POSSIBLE MECHANISMS OF CORONA DISCHARGE INVOLVED IN BIOGENESIS," NATURE VOL. 256, 1975 Pages 34-5.

BEAUFORT, W., ETL, "HEAVY METAL RELEASE FROM PLANTS INTO THE ATMOSPHERE," NATURE VOL. 256, July 3, 1975, Pages 35-7.

KAMA, A. K. AND AHIRE D. V., "ELECTRICAL ATOMIZATION OF WATER DRIPPING FROM PLANT LEAVES," VOL. 22, March 1983, Pages 509-511.

SIIRDE, ETAL, "ON SOME PECULIARITIES OF PLANT GROWTH UNDER ACTION OF IONIZATION," 9TH INTERNATIONAL CONGRESS OF BIOMETEOROLOGY, 1981.

FISH, B. R., "ELECTRICAL GENERATION OF NATURAL AEROSOLS FROM VEGETATION," SCIENCE, 175, 1239-1240.

VONNEGUT, B., AND NEUBAUER, R. L., "PRODUCTION OF MONODISPERSE LIQUID PARTICLES BY ELECTRICAL ATOMIZATION," JOURNAL OF COLLOID SCIENCE, 7, Pages 616-622.

DEROSA, ERNEST W., "LIGHTNING AND TREES," JOURNAL OF ARBORICULTURE, 9 (2), February 1983, Pages 51-3.

WENT, F. W., "PLANT LIFE," SCIENTIFIC AMERICA, 192:62, May 1955.

HARRINGTON, DANIEL B., ETAL, "EFFECTS OF SMALL AMOUNTS OF ELECTRIC CURRENT AT THE CELLULAR LEVEL," MARQUETTE UNIVERSITY, MILWAUKEE, WISCONSIN.

HELLMAN, HAL, "LIGHT AND ELECTRICITY IN THE ATMOSPHERE," HOLIDAY HOUSE, NEW YORK, 1968.

REDUCTION OF LIGHTNING ACTIVITY

Reduction of lightning has been a subject of research in weather modification for many years. This research, and subsequent experimental efforts, have been met with great controversy by the technical community. There was a research program entitled "Project Skyfire" that was initiated in 1953 and enjoyed great support with many governmental organizations and scientific foundations. The main thrust of the project was to modify weather. A part of that program was the effort to reduce lightning activity. This was to be accomplished by seeding clouds with material to reduce the charge build-up within the cloud mass. The program consisted of seeding clouds with ice crystals, dispensing silver-iodide smoke and/or long (i.e. 10 cm.), electrical conducting fibers in growing cumulus and/or developing cumuloumbus clouds.

There were some effects in both promoting the increase in lighting activity and decreasing the lightning activity. No decrease in lightning activity by seeding clouds was conclusively demonstrated. The lightning reduction program was terminated by the Federal government when apparent effects were shrouded in uncertainty. There was apparently some fear that the results of the lightning reduction effort in modifying weather might increase rather than decrease weather related damage, such as increasing the number of flood-producing storms. In an earlier weather modification program, seeding of a hurricane was attempted under Project Cirrus (1947). This hurricane suddenly changed course and moved across Savannah, Georgia, causing heavy damage. The seeding of Hurricane Ginger in 1971 was followed by damage in North Carolina. It might well be a noble product of mans continual search for modifying his environment to make the earth more habitable. But at what price and who will decide what is best for a given nation; there is no panacea. The benefits from modifying the weather may be obtained at the cost of other important secondary losses. The desire to change/or modify the weather patterns could have disastrous consequences.

Subtle changes in the amounts and/or locations of rainfall would produce many environmental and biotic changes. For example, the water level of streams and the increase in flow velocity of those streams would have a larger effect on organisms living in those streams than the superficial effects of any earth erosion. The flushing effect of the natural nutrients in lakes and other bodies of water could effect the balance of those levels of nutrients required for aquatic life. The changes in rainfall and sunshine due to thunderstorm modifications on vegetation could effect the life cycle of that vegetation. A small change in the amount of sunshine could cause a significant change in vegetation growth due to longer or shorter growing periods. These growth patterns are not linear effects, some vegetation would have thresholds where changes would result in the destruction or rapid growth. The results of growth of most vegetation would be different and we would expect to see some abrupt changes.

The effects of change in rainfall and sunshine on vegetation would be equally felt by animals having to adapt to the changing environment. The changes would not end here but would be extended to the smallest of the insect world up to and including mankind. During thunderstorm activity, the relative electrical forces due the charged atmosphere will be proportionally much higher. This indicates that lightning which produces and sustains the charged atmosphere, imparts to the sub-micron particles, the electrical motion which acts to precipitates out these fine particles onto cloud droplets which will ultimately fall back to the earth. If it were not for the charged atmosphere, these sub-micron particles would collect and stay suspended in our upper atmosphere producing climate changes which would eventually spell the end of life on earth.

The phenomena of lightning is not unique to the planet earth. NASA's Voyager I Planetary Explorer Spacecraft detected lightning embedded in the great storms that exist within Jupiter's atmosphere. This discovery was the first hard evidence that lightning takes place on other planets. Voyager 2, detected lightning discharges on Saturn and Uranus and the Pioneer Venus Orbiter picked up radio signals associated with lightning in the atmosphere of Venus. The detection of Lightning in atmospheres other than the earths' indicates that lightning is commonplace on other planets with atmospheres within our solar system. With these discoveries: that lightning discharges are occurring in the atmospheres of other planets tend to indicate that the development of strong electrification is not unique to our atmosphere but a general phenomena that can occur in a variety of gases and cloud particles that are very different from our atmosphere.

Life as we know it, is a planetary phenomena. Life on planet Earth had its origin intertwined with the interactions with liquid water, a gaseous atmosphere, and minerals provided within the solid planetary (i.e., Earth) surface. The energy required to produce the needed chemical reactions to produce life was available from solar ultraviolet light, charged particles, local volcanism, meteoritic impact, hydrothermal vents, lightning discharges, and acoustical shocks from thunder.

Lightning is a phenomena that exists within our solar system and most probability throughout the entire universe. These high energy electrical events (i.e., lightning) are truly natural phenomena which mankind cannot control nor regulate.

REFERENCES

CORLISS, WILLIAM R., "TORNADOS, DARK DAYS, ANOMALOUS PRECIPITATIONS AND RELATED WEATHER PHENOMENA," THE SOURCE BOOK PROJECT, GLEN ARM, MD, 1981

VONNRGUT, BERNARD, "POSSIBLE ROLE OF A ATMOSPHERIC ELECTRICITY IN THE DYNAMICS OF ATMOSPHERES MORE DENSE THAN THAT OF EARTH," A. DEEPAK PUBLISHING, HAMPTON, VA. 1983.

LOWRY, WILLIAM P., "THE CLIMATE OF CITIES," SCIENTIFIC AMERICA, 217,2 (1967).

MORRIS, EDWARD A., "THE LAW AND WEATHER MODIFICATION," BULL. AMERICAN METEOROLOGICAL SOCIETY, 46, 10, Pages 618-22, 1965.

STAFF, "LIGHTNING AND THE SPACE PROGRAM," KSC RELEASE NO. 72-90, JUNE, 1990.

LATHAM, JOHN, "PROCEEDINGS IN ATMOSPHERIC ELECTRICITY," A. DEEPAK PUBLISHING, HAMPTON, VA., 1983.

VONNEGUT, BERNARD, "POSSIBLE ROLE OF ATMOSPHERIC ELECTRICITY IN THE DYNAMICS OF ATMOSPHERES MORE DENSE THAN THAT OF EARTH," STATE UNIVERSITY OF NEW YORK, ALBANY, N.Y., 1980.

"THE SEARCH FOR LIFE'S ORIGINS," NATIONAL ACADEMY PRESS, WASHINGTON, D.C., 1990.

CONCLUSIONS

Lightning is a creation of nature, as such, it can be reasoned that it has a designated mission in assisting in sustaining the natural climatic order of the planet. This force can be fatal to man by charge transfer and/or lightning induced fires. We should not fear lightning but we should respect it. This book demonstrates that lightning is a friend to all living things. It is one of natures ways to keep in balance all of those forces necessary to sustain life on our planet. With each thunderstorm we experience there is an assurance that life is proceeding as planned and that life will continue.

Lightning is one of those great forces of nature that works harmoniously with the others to produce and sustain a living earth.

APPENDIX A Types of Lightning Dscharges/Flashes

There are three general types of lightning (Figure A-1). The most dominate type of lightning is the intra-cloud lightning, then there is cloud-to-cloud lightning and cloud to ground lightning. Nearly 60 to 70% of the lightning activity is in or between the clouds during thunderstorm activity. The base to the thunderstorm clouds are charged negative and the top portion of the cloud mass is charged positive. There is, on occasion, a small area of positive charge in the lower portion of the cloud mass. This normally occurs during the dissipation of the storm in the area of the precipitation downfall. The cloud to ground lightning normally brings down a negative charge to the earth. In mountainous land form regions positive charged lightning flashes are somewhat common. The lightning can be visually identified as negative or positive by the observing the pointing direction of the fingers of the flash. If the pointing direction of the fingers are downward (towards the earth) then the lightning flash is negative, if they point upwards (towards the clouds) then the flash is positive, this difference is shown in *Figure A2*.

A cloud to ground lightning flash normally consist of a number of lightning strokes. Normally there is three to four strokes in a flash and can have up to 15 or more strokes in a single flash. Lightning occurs when a sufficient charge has built-up, and separated, which overcomes the dielectric strength (i.e. resistance) of the adjacent air. The lightning starts with a electrical breakdown of the air. The forward progress towards the earth is in steps, pausing along the way, while the charge builds and the charge leader looks for easier (i.e., less resistance) path in its progress towards earth. This charge leader moves in steps of about 50 meters with a corresponding time of about 50 millionths of a second. This charge leader is called a stepped leader. This process of direction changing steps in the downward progress of the lightning streamer produces the jagged and forked design of the resultant lightning flash. As the stepped leader approaches the earth, the charge on the earth below increases rapidity in the opposite polarity. Ionized streamers are produced on the earth which travel upwards to meet the downward stepped leader from the charged clouds, when the two meet there is a brilliant illuminated exchange current between the earth and the charged clouds above; the result is the lightning flash.

This short discussion on the nature and cause of lightning is intended to give one a basic understanding of the mechanism(s) of lightning. We normally think of lightning as a cloud-to-ground phenomena, but there are several variations between clouds and the normal bright earth flash.

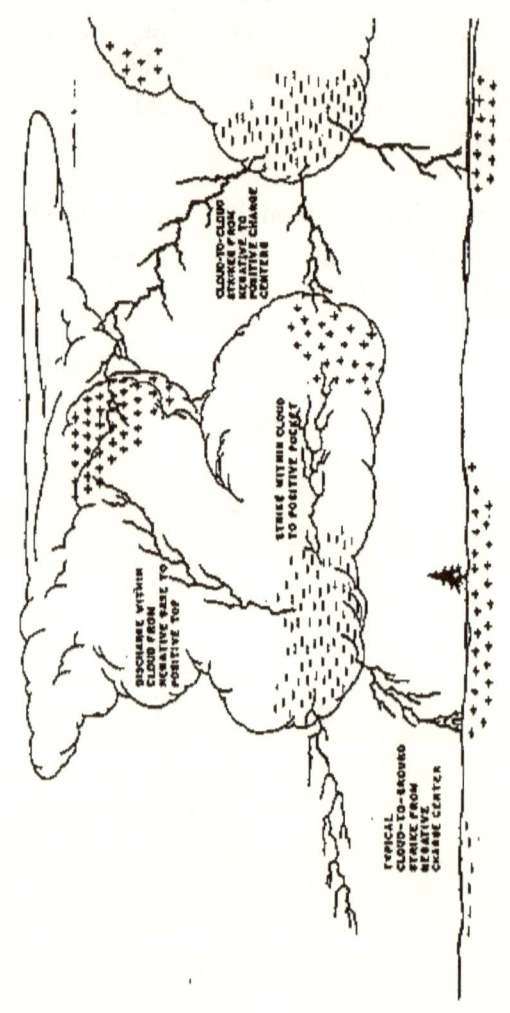

Figure A-1: Types Of Lightning Flashes

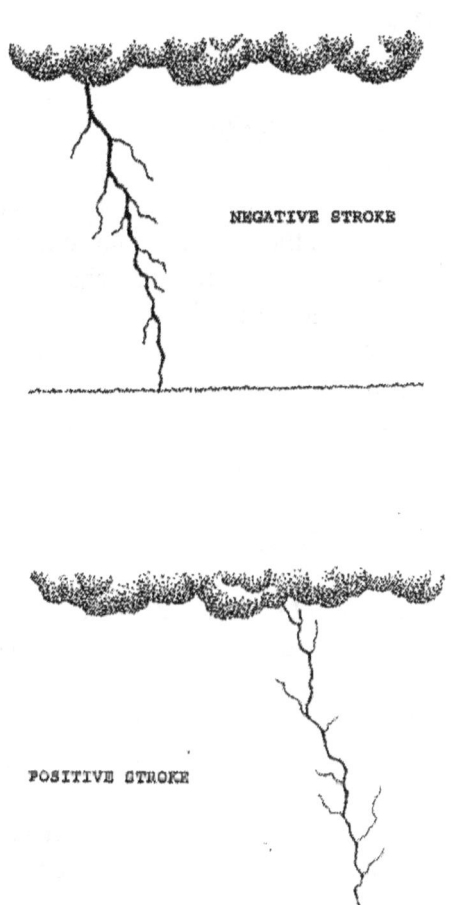

Figure A-2: Visual Difference Between a Negative and Positive Lightning Stroke

LIGHTNING'S DIFFERENT FLASH CONFIGURATIONS

Cloud-To-Ground Lightning

This type of lightning is the most spectacular of all the lightning types and the most familiar to most observers. This type produces a bright electrical discharge from the cloud(s) to earth. Associated with this brilliant flash is a loud audible sound which follows in time, the bright flash. The flash is composed of many bright branches and appears to "flicker". This flicker or pulsation is due to multiple strokes following the same discharge path in sub-second events.

Intra-Cloud Lightning

Intra-cloud lightning is the most common type of discharge which occurs between different parts of the cloud-within the cloud. This type of lightning discharges regions of the cloud with opposite charge. This entire process of discharge takes place within the cloud. Sometimes the brilliant flash will be seen at the edges of the cloud and tend to reveal a bright channel similar to cloud-to-ground lightning, and then re-enter the cloud formation.

Cloud-To-Cloud Lightning:

This type of lightning produces the bright flashes between clouds. The electrical charge of one cloud is influenced by the opposite charge of a nearby cloud or cloud formation. The clear air path between the clouds is breached by the lightning breakdown potential (voltage) is difference to the earth or intra-cloud charge centers.

Cloud-To-Air Lightning

This type of lightning is generally referred to as "Bolts from the Blue". This is due to this type of lightning being observed in blue sky miles from the cloud source. The flash exits the cloud and terminates in the clear air (In the Blue). This type is generally heavily branched, with each branch apparently ending in a region of space charge in clear air. Often such lightning strokes can meander for distances of up to 40 miles without striking any target finally dissipating. This type of lightning flash is comparatively rare and are more often seen in arid regions where the cloud bases are normally higher.

Heat Lightning

This type of lightning is normally intra-cloud lightning or lightning that lights up the cloud which is viewed by the observer at an angle that would hide a cloud-to-cloud discharge. This type of lightning is normally viewed at night and the cloud masses are so far from the observatory that they cannot hear the thunder associated with the lightning.

Sheet Lightning

This type of lightning is intra-cloud; the lightning produced a diffused illuminative which lights up a large portion of the cloud mass. The lightning is most common in thunderstorms where the cloud bases are so high above ground that it is easier for the discharge to pass from different charge centers within the cloud rather than the longer route to earth.

Ribbon Lightning

This type of lightning is applied to a flash that, to the naked eye, looks like several closely-spaced lightning strokes closely paralleling one another to ground. Actually the appearance results from the ionized path of a multiple-stroke lightning flash being blown sideways with the wind. The retina of the eye retains the image of all the strokes for a short time, and they seem to be occurring simultaneously.

Bead Lightning

A lightning flash in which the discharge channel appears to have alternate bright and faintly luminous, or nonluminous, sections. It, therefore, appears as a string of bright beads. This type usually begins as a normal discharge, with the entire channel bright. As the channel fades with time, some sections remain brighter than others. One explanation of this is that, because the channel is twisting all around, the observer is viewing different sections of the channel with a different perspective. If his line of sight is more or less perpendicular to the channel, he will not see very many luminous particles per unit area of his field of view, and that portion of the channel will be fairly dim. If he is looking along the channel, he will see more particles, and this section will appear bright. This phenomenon may also be related to a "magnetic pinch effect". Any electric current produces a magnetic field, which in turn tends to force the moving charges, which produce the field, closer together, thereby constricting the channel. If the current varies along the channel due to surges, some sections would be constricted more than others, presumably producing the bead-like visual appearance. The bead-lightning phenomenon apparently only occurs in conditions of restricted visibility, such as a heavy rainfall, when it is less likely that enough light from the initially bright channel will reach the observer's eye to saturate the retina.

Ball Lightning

This is possibly the most mysterious atmospheric electrical phenomenon. It appears as a luminous ball ranging in diameter from a few centimeters to a meter or so. Reports of its behavior vary considerably. Some seem to hover in midair; some travel at great speed. They sometimes bounce off of conductors, but sometimes appear to penetrate and remain inside of aircraft. Some last a few seconds; some persist for minutes. Some end their life in an explosion; some just fade away. In short, there appears to be no good physical explanation for a phenomenon with all of the reported properties of ball lightning. Scientists initially discounted reports of ball lightning, believing it to be a figment of the observer's imagination or a muddled description of some other optical phenomenon. It was later thought that ball lightning was real but rare. Some experts now believe that it is as common as cloud-to-ground lightning and may, in fact, be caused by it.

St. Elmo's Fire

This term, corona discharge, and point discharge are names given to the same physical process under different conditions. It is caused by the electrical field intensifying in the vicinity of a more or less pointed object. The charged particles moving in this field collide with and ionize some of the molecules in the area. This is an equilibrium process, and when some of the ions and electrons recombine, energy is released in the form of light. The air around the point glows, often with a very eerie effect. St. Elmo's Fire gets its name from the patron saint of sailors and was applied to the glow around mastheads during stormy conditions.

APPENDIX B: Lighting-Like Discharges

The thunderstorm source of lightning is the most common source but not the only source of lightning. Lightning-like discharges occur in sandstorms, snowstorms, in ejected material from erupting volcanoes, near the fire-ball created by nuclear explosions, in tornadoes, waterspouts, bolts from the blue and man-made triggered lightning. Normally, for lightning to occur, the atmosphere must acquire an electrical charge large enough to cause the electrical breakdown of air. There are other physical/electrical processes that can produce lightning-like discharges.

VOLCANIC INDUCED LIGHTNING

Lightning-like discharges are associated with volcanic eruptions. The electricity generated is associated with friction between the ash particles ejected and the surrounding gases. The convection and condensation of water vapor above the volcano during an eruption can charge the atmosphere very similar to that of the thunderstorm buildup. The larger ejected ash particles fall to the earth due to gravity while the smaller particles are carried upwards by the air currents and winds. The lighter-positively charged particles move upwards and the heavier negatively-charged move downward, producing a very large charge separation. The lightning associated with this volcanic activity is due to the electrical charge created by the large mass of volcanic ash/dust thrown into the air and the great speed in which it is ejected.

NUCLEAR INDUCED LIGHTNING

The lightning produced close to a nuclear fireball (See figure B-1) is initiated after the nuclear detonation. The lightning-like strokes are positive going strokes and are not multi-stroke flashes like the normal thunderstorm lightning flashes. The intense gamma-ray burst from the nuclear detonation strips electrons from the surrounding air molecules. These electrons move rapidly away from the detonation site creating a large charge separation. The lightning-like discharge is triggered at a point on the earth where the electrical field is very intense and the electrical discharge travels upwards through the region of the greatest negative charge producing the lightning-like discharge.

TORNADOES/WATERSPOUTS PRODUCED LIGHTNING

The powerful vortex motion of a tornado and a waterspout produce and intensify the electrical charge buildup within the storm which produces lightning-like discharges (See Figure B-2). The tornadoes's dynamics is responsible for the development of the charge buildup within the swirling vortex/funnel. Other electrical phenomena associated with the tornado consists of sheet lightning with differing brilliant colors surrounding the entire surface of the vortex. The vortex formation with its associated air movement creates friction between particles especially dust particles and attain an unusually large electrical charge intensity. The waterspout characteristics are essentially the same as the continental (i.e. land based) tornadoes. The main differences are: a. They have smaller dimensions, b. shorter durations, c. less electrical activity and, d. lesser strength.

DUST/SAND STORMS INDUCED LIGHTNING

Dust and sandstorms can have durations of several days and cover thousands of kilometers, spreading in width over hundreds of kilometers. Electrical arcing and lightning-like discharges are associated with the hot and dry dust/sandstorms. The electrical fields within these storms increase the charge on the dust/sand grains due to the friction with each other and with other particles. Whenever the dust or sand move, electrical charges and charging of the air originates. The dust/sand are negatively charged and the earth has a positive charge; this causes lightning-like discharging to the earth.

This same phenomena is associated with frictional charging of snow and dense snow or frozen particles. Intense electrical charge buildup(s) have been observed in severe wind driven snowstorms.

In the bolts from the blue and man-made triggered lightning; the region of the atmosphere associated with these phenomena must have an electrical charge sufficiently large enough to cause electrical breakdown of the air. The manmade triggered lightning is produced by firing rockets with a trailing copper wire into a charged cloud. The trailing copper wire provides a discharge path for the atmospheric electrical discharge to earth. Bolts from the blue are so-called because they appear from a blue sky with no clouds in view, actually the clouds are out of the view of the observer. Often such lightning strokes do not reach the earth and dissipate in the air after traveling 10's of miles in the atmosphere.

An unusual lightning-like discharge resulted from a depth charge set-off in the ocean by the U.S. Navy which sent a plumb of water upwards (See Figure B-3). This plumb of water triggered a lightning-like discharge from the clouds above to the water column.

In foregoing as described above, natural physical mechanisms provided for the transfer of the dynamically produced electrical energy to other forms of energy. In each case, that transfer of electrical energy was in the form of lightning-like discharges.

Figure B-1 Lightning Induced by a Thermonuclear Explosion

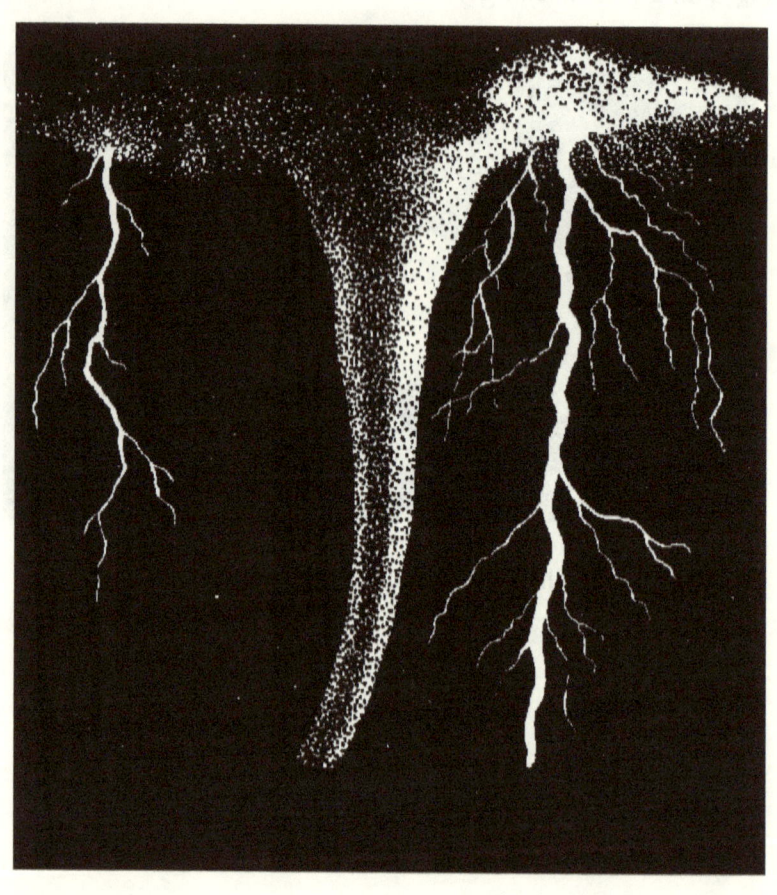

Figure B-2 Lightning Activity Associated With A Tornado

Figure B-3 Lightning Produced by an Underwater Explosion

APPENDIX C: The Signature of Lightning

To complete the discussion of Lightning flashes and their different configurations, a review of the Lightning's color and final configuration as it enters the earth and deposits its energy is discussed.

The lightning flash normally appears as brilliant white in color. This white appearance is actually composed of many colors. The absorbed energy from the lightning flash causes the nitrogen and oxygen atoms in their atmospheric path to be ionized and radiate visible light. When water vapor in the atmospheric path of the lightning flash is ionized, the reddish or pinkish in color. The forest ranger relates the occurrence of forest fires to white lightning but not so with the reddish color lightning flash which is associated with the ionization of rain water in the lightning path. Other such observations associate blush color lightning with hail storms and storms that begin with red color lightning end with yellowish lightning flashes. White lightning is normally associated with a single loud thunder report while reddish lightning is followed by rolling thunder. The color as perceived by the observer of the lightning flash can be dependent on the distance away, cloud cover, and other atmospheric environmental conditions. Red and yellow flashes are often associated with air that is heavily charged with dust. On rare occasions violet and green flashes have been reported. Blue flashes are normally associated with exceptionally violent thunderstorms.

In a recent research program, the researchers (University of Alaska) recorded fleeting flashes of red light associated with high altitude electrical storms above South America. This same type of phenomena, that of similar red and blue burst of light high in the atmosphere above the cloud cover associated with large developed thunderstorms was observed in aircraft observations in the Midwest USA.

In the flights over South America, there was no observations of the "blue" burst of light which were associated with the larger thunderstorms in the Midwest USA. The term for these high altitude burst of light associated with thunderstorms is "SPRITES". The origin of the use of the term sprite suggests an imaginary being, pleasing appearance, to be admired for their lightness of movement, they may, however be mischievous or even hostile. The term appeared in published accounts of a NASA sponsored program conducted in June and July, 1994. They reported flashes that looked like the Forth of July; blue or purple hue jets appear in narrow beams, spray patterns, fans, and/or cones of light that appear to originate at the top of the cloud formation and extend upwards to altitudes of about 20 miles. These jets of light are ejected in all upward angles at speeds of 20 to 60 miles a second.

As magnificent and specular as lightning is, the only lasting record is the fused imprint of the stroke as it deposits its energy in the sandy and rocky formation of the earth's sub-surface. When lightning strikes it violently disturbs the sand and rock within its path and can burn a channel within the earth's subsurface radiating outwardly producing fingered furrows. If the lightning energy is conducted through sandy and some forms of rocky soil, the lightning current can cause a change in state of the material in its path. A metamorphosis occurs which freezes this lightning path in a geometrical design similar to that of a lightning stroke as it passes through the atmosphere to the earth. The lightning current flow can form tubes of fused silica (melting point of 1,710 degrees centigrade). As the lightning enters the earth, it produces a central hole which resembles a Molehole, with outward radiating structures. These structures (tubes) of fused glassy type material are formed along the discharge path. These structures can range in size from 1/2 to 3 inches in diameter and extend up to 40 feet in length. These resulting structures are called "Fulgurite" (A glassified tube formed in sand or rock by lightning) and represent a lasting permanent record of the Lightning occurrence: "The Signature of Lightning." The Fulgurites have been found at depths of approximately 60 feet but many of the other Lightning paths are very shallow, with a resulting lightning signature of scorched grass in a kaleidogram appearance. Of those fossilized lightning signatures unearthed; there are those that were found to have been formed nearly 250 million years ago.

APPENDIX D: SUMMARY GUIDE FOR PERSONAL SAFETY DURINGTHUNDERSTORMS

PERSONAL SAFETY DURING THUNDERSTORMS

Most of the fatalities due to thunderstorms could be avoided. Thunderstorms do not come without warning, there is nothing "sudden" about thunderstorm formations. The best protection against being caught in an exposed area during a thunderstorm is to be able to recognize the cloud formations preceding the storm. Keep a watchful eye on cumulus clouds that darken and thicken rapidly, especially during the afternoon. When planning an outdoor activity, make sure that your planning includes getting an updated weather forecast.

BE WATCHFUL FOR THUNDERSTORM ACTIVITY

Most of the thunderstorms within the United States occur in the early to late afternoon. In the southeastern states, less than 20% of the summer thunderstorms occur between 6 PM and midnight. In at least fifty percent of the states, half of the summer thunderstorms occur between the hours of noon and 6 PM.

Meteorologist can with great accuracy forecast the movement of the frontal thunderstorm activity. Their forecast is based on charting the movement of the warm and cold air masses as they travel across the country producing thunderstorm climatic conditions. This type of frontal thunderstorm activity may last for hours, with heavy rains, high winds and lightning activity. Once the front has passed through, the stormy weather ceases and the skies begin to clear quickly.

The local heat-type generated thunderstorms are more difficult to forecast. They develop on hot and humid summer afternoons. The process normally begins with a darkening and build up of the cloud cover. As the day progresses, the clouds gradually evolve into Cumulonimbus clouds. There will be heard the distant peal of thunder preceding the darkening of the overhead sky. The local type thunderstorms do not last long, normally about 1/2 to one hour in length. These storms can produce some gusty winds and brief and gushing rains with hail and lightning.

The large and severe thunderstorms are those that are produced by the fast moving cold fronts. They can produce heavy rainfall, high velocity winds and extended an dangerous lightning activity. With today's technology, it is quite easy to keep up with weather forecast. Every radio station, television station, and newspaper provides the weather forecast. Before planning outdoor activities, the weather forecast should be reviewed and plans made accordingly.

CLOUD FORMATIONS ASSOCIATED WITH THUNDERSTORM ACTIVITY

The cumulonimbus cloud formations as shown in *Figure D-1* are a prerequisite to the development of a thunderstorm. Its towering majestic form can reach heights of 40,000 feet or more. These clouds develop in a vertical upward moving formation as a result of strong upward air currents, which can be created in a number of ways. Local land mass heating can create the upward rising air motions by warming the air near the surface of the earth. This warm air will rise forming a cloud overhead, or a much larger disturbance of a cold air front moving through the area forcing the air in its path upward. Along the coast or areas adjacent to large water areas, there can exist a temperature differential between land and the water which can induce the upward rising motion of cloud forming air masses.

Figure D-1. Cumulonimbus Cloud Formations

The air cools as it rises and moisture within the air may condense. Small, puffy Cumulus clouds as shown in *Figure D-2* maybe be the first clouds to be formed. If the atmosphere is unstable, the air within the Cumulus will continue to rise and produce more water vapor due to the cooler temperatures, which will increase the cloud cover and develop larger cloud masses. When these cloud masses darken and grow rapidly with a increase in the upper atmospheric and surface winds; a thunderstorm could develop. The cumulus clouds may develop into Cumulonimbus cloud formation. The Cumulonimbus clouds exhibit great vertical development. The cloud tops can spread out and form an anvil configuration. To an observer directly beneath, the cumulonimbus clouds may cover the entire sky and have the appearance of Nimbostratus (see *Figure D-3*). Many times, preceding the forward movement of this cloud mass, there can be violent lightning discharges.

Some of the lightning activity may be well ahead of the darker cloud mass and strike from less ominous looking clouds. Because of the obvious relation of cloud cover, increases in vertical and/or surface winds, the approach of a thunderstorm can be easily identified. The direction of the moving storm can be visually observed at a distance by watching the movement of the clouds. These storm fronts can move very fast an require that someone in the outdoors to seek shelter in a timely manner.

Figure D-2. Cumulous Cloud Formations

Figure D-3 Nimbostratus Cloud Formations

PERSONAL CONDUCT

This guide is intended to provide a summary of information for personal safety during a thunderstorm.

a. Do not go out-of-doors or remain out during thunderstorms unless it is necessary. Seek shelter as follows:
    1. Dwellings or other buildings which are protected against lightning.
    2. Underground shelters such as: subways, tunnels, caves, earthen embankments.
    3. Large metal-frame buildings.
    4. Enclosed automobiles, buses and other vehicles with metal tops and bodies.
    5. Enclosed metal trains and street cars.
    6. Enclosed metal boats or ships.

7. Boats which are protected against lightning.
8. City streets which may be shielded by nearby buildings.

b. If possible, avoid the following places which offer little or no protection from lightning.
1. Small unprotected buildings, barns, sheds, etc.
2. Tents and temporary shelters.
3 Automobiles (non-metal or open).

Certain locations are extremely hazardous during thunderstorms and should be avoided if at all possible. Approaching thunderstorms should be anticipated and the following locations avoided when storms are in the immediate vicinity.

a. Hilltops and ridges.
b. Areas on top of buildings.
c. Open fields, athletic fields, golf courses.
d. Parking lots and tennis courts.
e. Swimming pools, lakes, fishing areas and seashores.
f. Near wire fences, clotheslines, overhead wires and railroad tracks.
g. Under isolated trees.

In the above locations, it is especially hazardous to be riding in any of the following during lightning storms.

a. Open tractors or other farm machinery operated in open fields.
b. Golf carts, scooters, bicycles or motorcycles.
c. Open boats (with or without masts) and hover crafts.
d. Automobiles (non-metal top or open).

It may not always be possible to choose a location that

offers good protection from lightning. Follow these rules when there is a choice in selecting locations:

    a. Seek depressed areas (low land farm areas) - avoid hilltops and high places.
    b. Seek dense woods - avoid isolated trees.
    c. Seek buildings, tents and shelters in low areas - avoid unprotected buildings and shelters in high (elevated) areas.
    d. If you are hopelessly isolated in an exposed area and you feel your hair stand on end, indicating that lightning is about to strike. Crouch down with your arms wrapped around your legs making a low profile. Do not lie flat on the ground or place your hands on the ground while kneeling down.

Appendix E: Glossary

Absorption: The process in which incident radiant energy is retained by a substance.

Aerosol: A colloidal system in which the dispersed phase is composed of either solid or liquid particles, and in which the dispersion medium is some gas, air or fog (aer, air; sol, from solution).

Air: The mixture of gases comprising the Earth's atmosphere.

Ampere: The electrocurrent's strength, or current density that one volt can send through a copper wire having a resistance of one volt.

Atmosphere: The envelope of air surrounding the earth and bound to it more or less permanently by virtue of the Earth's gravitational attraction.

Atmospheric Pressure: The pressure exerted by the atmosphere as a consequence of gravitational attraction exerted on the "column" of air lying directly above the point in question.

Angstrom: A unit of length used in the measurement of electromagnetic radiation and in the measurement of molecular and atomic diameters. One angstrom equals 10 to the power of minus eight centimeters.

Cirrus: A principal cloud type (cloud genus) composed of detached cloud elements in the form of white, delicate filaments, of white (or mostly white) patches, or of narrow bands.

Cloud: A visible collection of particles of water or ice suspended in the air, usually at an elevation above the earth's surface.

Climate: "The synthesis of the weather"; the long term manifestations of weather, however they may be expressed.

Cloud Height: In weather observations, the height of the cloud base above the local terrain.

Cloud Mass: That portion of the sky cover which is also called cloudiness or cloudage.

Cloud Seeding: Any technique carried out with the intent of adding to a cloud certain particles that will alter the natural development of that cloud.

Condensation: The physical process by which a vapor becomes a liquid or solid: the opposite of evaporation.

Cosmic Rays: A stream of atomic nuclei of heterogeneous extremely penetrating character that enter the earth's atmosphere from outer space at speeds approaching that of light and bombard atmospheric atoms producing secondary particles.

Cumulonimbus: A principal cloud type (cloud genus), exceptionally dense and vertically developed, occurring either as isolated clouds or as s line or wall of clouds with separated upper portions. These clouds appear as mountains or huge towers, with at least a part of the upper portions usually appearing smooth, fibrous, or striated, and almost flattened. This part often spreads out in the form of an anvil or vast plume.

Cumulus: Detached clouds simulating rising mounds, domes or towers, the upper part resembling a cauliflower (Cumulus, heap).

Direct Current (DC): A relative steady current in one direction in an electric circuit, which creates a continuous stream of electrons through a conductor.

Downwind: The direction toward which the wind is blowing; with the wind.

Ecology: A branch of science concerned with the interrelationship of organisms and their environment.

Electric Field (E): The force at a point in space which would be experienced by a unit electric charge. The gradient (space rate of change) of the work required to move a unit electric charge from one position to a nearby position.

Electric Field Strength: The quantity of volts per meter.

Electrosphere: The Electrosphere is located some 50 kilometers above the surface of the earth. This is an ionized region below the Ionosphere.

Evaporation: The physical process by which a liquid or solid is transformed to a gaseous state; the opposite of condensation. (Evaporation is also called vaporization)

Fair Weather (Fine-Weather): The fair or fine weather describes the electrical state of the atmosphere which is free of clouds, rain or thunderstorms. Below an altitude of a few tens of kilometers, there is a downward-directed electric field during fair weather of 120 to 130 volts per meter.

Forked Lightning: A common form of a cloud-to-ground discharge, that exhibits downward-directed branches from the main lightning channel.

Front: The transition zone between two air masses of different density and temperature, typical for weather changes.

Fulgurant: Flashing like lightning.

Fulgurate: To destroy (as in abnormal growth) by electricity.

Fulgurite: A glassified tube formed in sand or rock by lightning.

Fulrurous: Lightning-like.

Greenhouse Effect: The heating effect created by the atmosphere upon the earth by virtue of the fact that the humid atmosphere absorbs and remits infrared heat radiation. Shorter electromagnetic radiation wavelengths are transmitted rather freely through the atmosphere to be absorbed at the earth's surface. The Earth then remits this as longwave (infrared) terrestrial radiation, a portion of which is absorbed by the atmosphere and again emitted.

Hertz (Hz): One cycle (c) per second (s), i.e.

Hz = 1 c/s

Ion: Electrically charged atoms or molecules that create an electrically conducting medium. They move quickly in the atmosphere (ion, going).

Ionization: In atmospheric electricity, the process by which neutral atmospheric molecules or other suspended particles are rendered electrically charged chiefly by collisions with high-energy particles.

Ionosphere: Portion of the earth's atmosphere, extending upwards from about 60 Km to an indefinite height, which is characterized by a concentration of ions and free electrons high enough to cause reflections of radio waves.

Lightning Channel: The irregular path through the air which a lightning discharge occurs.

Lightning Discharge: The series of electrical processes by which charge is transferred along a channel of high ion density between electric charge centers of opposite sign (1) within a thundercloud (cloud discharge), (2) in a cloud and on to the Earth's surface (cloud-to-ground), (3) within two different clouds (cloud-to-cloud discharge), or (4) in a cloud and in the air below the cloud (air discharge).

Lightning Flash: In atmospheric electricity, the total observed luminous phenomenon accompanying a lightning discharge. A flash is composed of a number of lightning strokes.

Lightning Stroke: Any one of a series of repeated electrical discharges comprising a single lightning discharge; specifically, in the case of the cloud-to-ground discharge, a stepped leader plus its subsequent return streamer.

Nimbostratus: Cloud in the form of a rainy layer.
Ozone: A nearly colorless (but faintly blue) gaseous form of oxygen, with a characteristic odor like that of weak chlorine. Its formula is O_3; its molecular weight is 48.

Point-Discharge: Visible and audible corona produced under the influence of thunderclouds.

Polarization: The action or process of affecting light or other radiation so that the vibrations of the wave assume a definite form.

Potential Gradient This is the change in voltage per unit length (distance) from the

(Voltage Difference):	surface of the Earth up into the charged atmosphere.
Smog:	Smoke-fog, a fog in which smoke or atmospheric pollutants combine an aerosol (i.e. a dispersion of small particles in foggy air).
Stratosphere:	That region of the atmosphere, lying above the Troposphere, it extends to a height of about 50 Km ("Stratos," layer).
Thunderstorm:	A storm accompanied by lightning and thunder.
Troposphere:	The lower layer of the atmosphere extending to 9-16 Km above the earth surface. In this region, the decisive changes of temperature, precipitation, lightning and cloud formations take place, ("Tropos," turning point).
Ultraviolet Radiation:	(Abbreviated as UV), electromagnetic radiation of shorter wavelength than visible radiation but longer than x-rays; roughly in the wavelength range of 10 to 4000 Angstroms.
Voltage (V):	The electromotive force or electric tension, also called electric potential difference.

Watt (W): The power of a current of 1 ampere flowing across a potential difference of one volt.

UNITS OF MEASURE

The following conversions factors make it possible to change from the Metric System to the English System. The abbreviations used in this book are given in parentheses. Note that the number 10 with a superscript means a number 1 followed by a number of zeros equal to the superscript. For example; 10 x5 means 100,000 and 10 x3 means 1000. When the superscript is negative, the number is 1 over 10 to the same power. For example 10 x^{-5} means $1/10\text{x}^5$ or 1/100,000.

1. **LENGTH:**

1 Meter (m)	39.37 inches (in.)
1 centimeter (cm)	0.3937 inches (in.)
1 Micron (U)	= 10x-4 Centimeters (cm)
	= 0.00003937 inches (in.)
1 kilometer (km)	= 10 x 3 Meters (m)
	= 3280.84 feet (Ft.)
	= 0.6214 Statute Mile (Mi.)
1 Statute Mile (Mi)	= 5280 Feet (Ft)
	= 1.0693 kilometers (km)

89

2. **TIME:**

Microsecond (Usec)	=	10^{-6}
Millisecond (ms)	=	10^{-3}
Nanosecond (ns)	=	10^{-9}

3. **VELOCITY:**

1 Mile per hour (MPH)	=	1.4667 Feet per Second (FPS) (Ft/sec)
	=	0.4470 Meters per Second (MPS)
1 Meter per Second (MPS)	=	2.2369 Miles per Hour (MPH)
	=	3.2808 Feet per Second (FPS) (Ft/sec)

APPENDIX F: OTHER SUGGESTED READING

1. BYERS, H. R., "THUNDERSTORM ELECTRICITY," UNIVERSITY OF CHICAGO PRESS, 1953.

2. MARSHALL, J. L., "LIGHTNING PROTECTION," JOHN WILEY & SONS, N.Y., N.Y., 1969.

3. UMAN, MARTIN A., "LIGHTNING," McGRAW HILL BOOK COMPANY, N.Y., N.Y., 1969.

4. UMAN, MARTIN A., "ALL ABOUT LIGHTNING," DOVER PUBLICATIONS, INC., NEW YORK, 1986.

5. BLANCHARD, DUNCAN C., "FROM RAINDROPS TO VOLCANOES," DOUBLEDAY & COMPANY, INC., GARDEN CITY, NEW YORK, 1967.

6. SMITH, L. G., "RECENT ADVANCES IN ATMOSPHERIC ELECTRICITY," PERGAMON PRESS, N.Y., N.Y., 1958.

7. CHALMERS, J. A., "ATMOSPHERIC ELECTRICITY," PERGAMON PRESS, LONDON, 1967.

8. HELLMAN, HAL, "LIGHT AND ELECTRICITY IN THE ATMOSPHERE," HOLIDAY HOUSE, INC., NEW YORK, 1967.

9. MALAN, D. J., "PHYSICS OF LIGHTNING," THE ENGLISH UNIVERSITY PRESS LTD, LONDON, 1963.

10. BRANCAZIO, PETER J. AND CAMERON, A. G. W., "THE ORIGIN AND EVOLUTION OF ATMOSPHERES AND OCEANS," JOHN WILEY & SONS, NEW YORK, 1964.

11. GOLDE, R. H., "LIGHTNING," VOLUMES I & II, ACADEMIC PRESS, NEW YORK, 1977.

12. SCHONLAND, BASIL, "THE FLIGHT OF THE THUNDERBOLTS," CLARENDON PRESS, OXFORD, 1964.

13. CORNITI, SAMUEL C. AND HUGES, JAMES, "PLANETARY ELECTRODYNAMICS," GORDON AND BREACH SCIENCE PUBLISHERS, N.Y., 1969.

14. WALLACE, JOHN M. AND HOBBS, PETER V., "ATMOSPHERIC SCIENCE," ACADEMIC PRESS, NEW YORK, 1977.

15. NEIBURGER, MORRIS, *ET AL*, "UNDERSTANDING OUR ATMOSPHERIC ENVIRONMENT," W. H. FREEMAN AND COMPANY, SAN FRANCISCO, CA , 1973.

16. PARKER, SYBIL P., "METEOROLOGY SOURCE BOOK," McGRAW-HILL BOOK COMPANY, NEW YORK, 1987.